Power Electronics
for the
Microprocessor Age

Power Electronics
for the
Microprocessor Age

Takashi Kenjo

Professor in the Department of Electrical Engineering
and Power Electronics
Polytechnic University of Japan

OXFORD NEW YORK TOKYO
OXFORD UNIVERSITY PRESS
1994

Oxford University Press, Walton Street, Oxford OX2 6DP
Oxford   New York   Toronto
Delhi   Bombay   Calcutta   Madras   Karachi
Kuala Lumpur   Singapore   Hong Kong   Tokyo
Nairobi   Dar es Salaam   Cape Town
Melbourne   Auckland   Madrid
and associated companies in
Berlin   Ibadan

Oxford is a trade mark of Oxford University Press

Published in the United States
by Oxford University Press Inc., New York

© Takashi Kenjo 1990
First published in paperback (with corrections) 1994

British Library Cataloguing in Publication Data
Kenjo, Takashi
Power electronics for the microprocessor age
1. Electronic motors. Control devices
I. Title
621.46'2

Library of Congress Cataloging in Publication Data
Kenjo, Takashi,
Power electronics for the microprocessor age/Takashi Kenjo.
Includes bibliographies and index.
1. Power electronics. 2. Electric driving. I. Title.
TK7881.15.K46 1989 621.46'2—dc19 88–27279
ISBN 0 19 856508 9 (Pbk)

Printed in Great Britain by
Courier International Ltd,
Tiptree, Essex

Printed and bound in Great Britain
by Biddles Ltd, Guildford and King's Lynn

Preface

In March 1983 in Tokyo, I attended an international conference on Power Electronics and heard the opening lecture by renowned Professor R. G. Hoft. He remarked that the advent of the microprocessor was changing the technology of this area. In some papers read in the sessions I sensed the quickening tide of a trend indicating that, for driving motors, bipolar transistors and new devices like MOSFETs would become the main power devices rather than conventional thyristors. In that atmosphere I was inspired to write this book.

However, I could not start it immediately because I had already begun revision and translation of my second English-language book *Permanent-magnet and brushless DC motors*. The first one, *Stepping motors and their microprocessor controls*, which had been written two years before, was published the next year. Even after these had been completed I was still busy in creating our KENTAC mechatronics controllers and related pieces of educational equipment, and also in writing three books in my mother tongue for Japanese and nearby Asian engineers. I believe that these pieces of work consolidated the firm basis on which this international version has been built.

This volume covers various aspects of microprocessor-controlled power electronic drives of electrical machines, focusing particularly on control-use motors of low and medium power. The general concept of this subject is discussed in Chapter 1, with some simple examples of microprocessor software for operation of stepping motors and a.c. motors. In the following chapter, various types of solid-state devices, their characteristics and basic circuit configurations are surveyed. Chapter 3 focuses on the a.c.-to-d.c. converters and phase controllers. Chapter 4 deals with the principles and theory of d.c. converters. One application of the d.c. converter is the servo-amplifier used for driving a d.c. motor in speed and position controls. This is discussed in the following chapter.

Spectacular development has been seen in miniaturization of stepping motors and their microprocessor controls since the appearance of my above-mentioned book. Chapter 6 is included for interpretation and some details of these technologies. Chapter 7 is a big chapter dealing with inverters or d.c.-to-a.c. converters for controlling the motion of a.c. motors. Certain developments have also been made in the area of brushless d.c. motors and their electronics controls. These are dealt with in Chapter 8.

To elaborate knowledge of possibilities and limitations of solid-state power electronics, Chapter 9 presents general theory on power conversion implemented through a bridge circuit consisting of four power devices. In the final chapter, a fundamental but useful concept of employing a microprocessor for position control using a d.c. motor is presented with a detailed explanation of software.

I would like to acknowledge that this book was completed with the friendly support and encouragement of several persons. Mr H. Stanbury, senior editor, advised me to produce this book to complete a trio with the above two books and was very patient in awaiting the final typescript. Mr T. Kawawaki, manager of the Tokyo Branch of Oxford University Press, kindly arranged encouraging circumstances for producing the book. Mr T. Takano, president of Sogo Electronics Publishing Company, helped me in producing 300 excellent illustrations. I wish to express my sincere gratitude to these persons. My thanks are also extended to my colleagues and students, who have collaborated effectively in the research and development that contributed to producing this volume.

Tokyo, Japan
October 1989

T. K.

Contents

1 Computers, power electronics, and motors

In most numerically controlled equipment or robots, a number of electric motors are used and their motion is governed by some form of artificial intelligence using microcomputers. However, between a microcomputer and a motor there is a very important stage, comprising solid-state power devices such as transistors, that receives control signals from the computer and deals

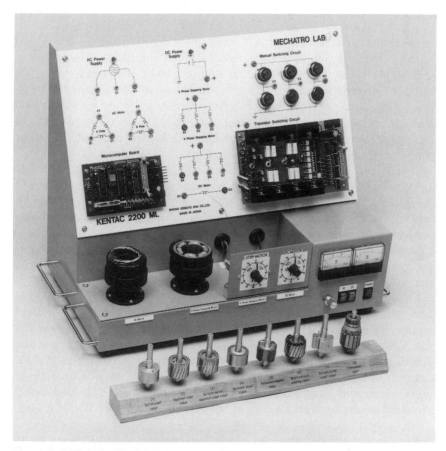

Fig. 1.1. MECHATRO LAB: an experiment bench designed to serve for the study of the relation between computer, power electronics, and motors.

with the electric power to be supplied to the motor to control its motion. This part of power electronics is the major subject of this book. In this chapter, the elementary relation between computers, power electronics, and motors will be explained, with reference to a teaching aid designed by the author. Since References [1] and [2] discuss in detail stepping motors and conventional and brushless d.c. motors, the fundamental characteristics of these motors will not be described here; however, some of the characteristics of a.c. motors will be surveyed.

1.1 MECHATRO LAB—an experiment bench

To help beginners to grasp the fundamental relations between a computer, power electronics circuits, and motors, the author designed the experiment bench that is shown in the photograph of Fig. 1.1. This experiment instrument, called MECHATRO LAB, has a single-board computer, several electrical circuits, electronic circuits, two stators (A and B) and several rotors. The single-board computer has an eight-bit 8085 microprocessor as its CPU, to govern the electronic circuits. As indicated in the system block diagram of Fig. 1.2, MECHATRO LAB provides a d.c. power supply and three-phase supply. It is also seen that stator A has two sets of conventional three-phase windings of four-poled and eight-poled arrangement. The other stator has three-phase concentrated windings on its six teeth (see Fig. 1.5).

Details of the single-board computer employed in the MECHATRO LAB are given in Fig. 1.3. It is seen that this compact computer consists of three major LSI (large-scale-integrated circuits): microprocessor (8085A), ROM (read-only memory) chip 2764 storing software, and a chip containing RAM (random-access memory) and I/O (input/output) ports. It is possible to drive the power electronics circuit from an external controller. In this book, some software examples will be presented for an external controller that is assumed to have a Z80 microprocessor.

Figure 1.4(a) is a photograph of the transistor power circuit, consisting of four pairs of cascade-connected transistors, and which can be used for a number of purposes. Figure 1.4(b) is the power-stage circuit diagram and Fig. 1.4(c) is the portion for interfacing to the output port of the computer. Application details of this circuit will be dealt with at various places in other chapters. It should, however, be emphasized that some primitive functions of power electronics can be demonstrated by the mechanical switching circuit mounted on the right above part of the main board of the MECHATRO LAB.

The mechanical construction of the two stators A and B is shown in Fig. 1.5. Stator A can be used as the stator of various types of a.c. motors and the brushless d.c. motor, while stator B may be used also as another stator of

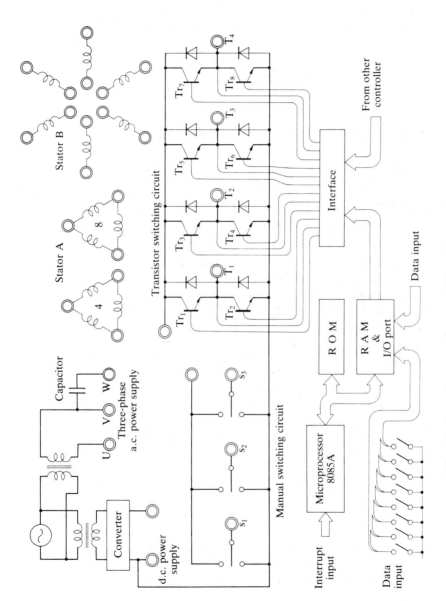

Fig. 1.2. Overall system diagram of the MECHATRO LAB.

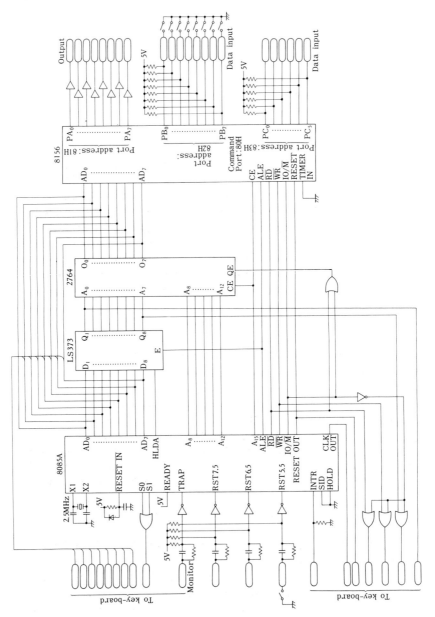

Fig. 1.3. Details of the MECHATRO LAB computer.

these machines, but is more suited to a stepping motor or a d.c. motor. The rotors are illustrated in Fig. 1.6, and are as follows:

(1) solid-steel rotor;
(2) squirrel-cage rotor;
(3) salient-poled squirrel-cage rotor;
(4) semihard steel rotor;
(5) permanent-magnet rotor;
(6) short-circuit winding rotor;
(7) salient-poled solid-steel rotor; and
(8) commutator rotor.

By combining one of the two stators, one of the eight rotors, the electronic circuit, and a computer software, we can develop an experimental system for power electronic control of a motor. For example, when stator A or B is used for constructing a brushless d.c. motor in the combination of the Hall elements shown in Fig. 1.7 for the rotor-position sensor, one can select any of rotors 3, 5, and 7. By changing the combination of these and by selection of the transistors in the power circuit to be used, we can investigate a large number of different ideas. Moreover, software flexibility enriches variation.

The purpose of this book is to study the principles of power electronic circuits, methods of driving a motor, and applications of microprocessors in power electronics.

1.2 Microprocessor-controlled power electronics drive of a variable-reluctance stepping motor

For an example of the relationship between the microprocessor, power electronics, and motor, let us first consider a three-phase variable-reluctance stepping motor, and operate it using the d.c. power supply and the mechanical circuit of Fig. 1.2. A variable reluctance motor[1] is composed of stator B and the salient-poled steel rotor No. 7. The connection of the windings and switches should be arranged as shown in Fig. 1.8(a). It should be noted that the coils are connected to form a two-poled three-phase scheme and that only three switches are used to commutate current.

We shall first study how this machine works, using Fig. 1.8(b), which is equivalent to Fig. 1.8(a) but more convenient in connection with the following explanation. The stator core has six salient poles or teeth, while the rotor has four poles. Both stator and rotor cores are made of mild steel having high permeability. Current is supplied from a d.c. power source to the windings via three switches, which are indicated by S_1, S_2, and S_3. In state (1), the winding of Phase 1 is supplied with current through S_1; we say 'phase 1 is energized or excited' in technical terms.

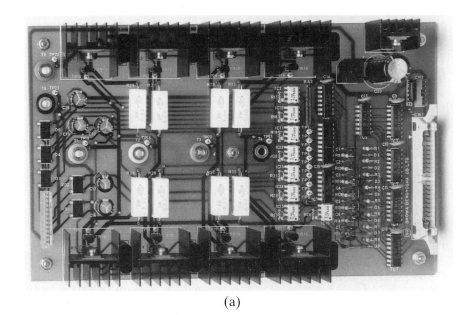

(a)

(b)

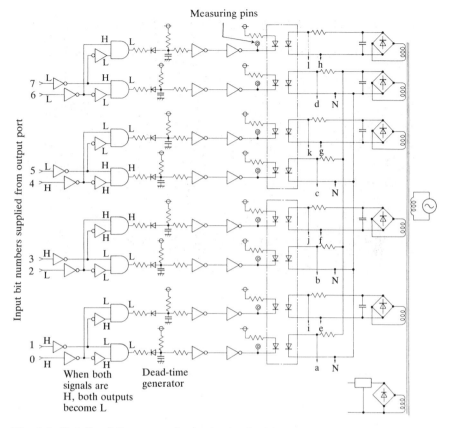

Fig. 1.4. Details of the power electronic circuit: (a) overall view; (b) power stage; (c) interface stage.

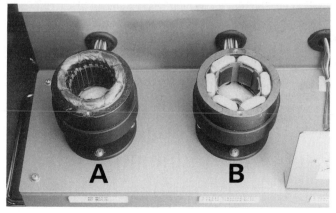

Fig. 1.5. Stators used in the MECHATRO LAB: stator A is for an a.c. motor and brushless d.c. motor, and stator B for variable-reluctance stepping motor and a conventional d.c. motor.

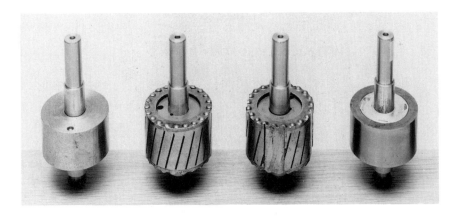

(1)	(2)	(3)	(4)
Solid-steel	Squirrel-cage	Salient-poled	Semihard steel
rotor	rotor	squirrel-cage rotor	rotor

(5)	(6)		(7)	(8)
Permanent-magnet	Short-circuit		Salient-poled	Commutator
rotor	winding rotor	(a)	steel rotor	rotor

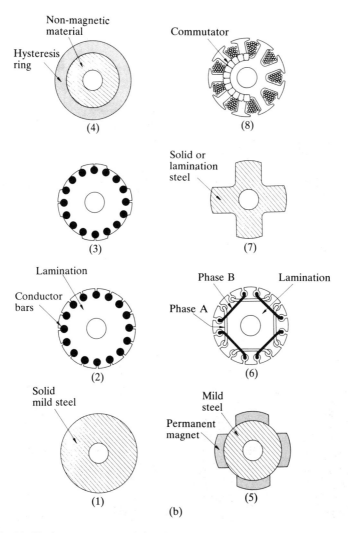

Fig. 1.6. (a) Various rotors used in the MECHATRO LAB; (b) their cross-sectional structures.

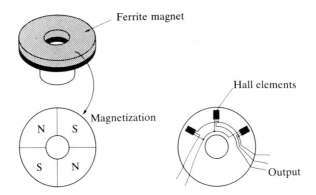

Fig. 1.7. Position sensors using Hall elements.

The magnetic flux that occurs in the air-gap between the stator and rotor owing to this excitation is indicated by a cluster of short lines. In this state, the two teeth of Phase 1 being excited (energized) are in alignment with two of the four rotor teeth. When S_2 is closed to excite Phase 2 in addition to Phase 1, magnetic flux is built at the stator teeth of Phase 2 in the manner shown in state (2), and a counter-clockwise torque is created owing to 'tension' in the inclined magnetic field lines. The rotor will then, eventually, reach state (3).

Thus, as one switching operation is carried out, the rotor rotates through a fixed angle, which is termed the step angle and is 15° in this case. If S_1 is now opened to de-energize Phase 1, the rotor will travel another 15° to reach state (4). The angular position of the rotor can thus be controlled in units of the step angle by a switching process. If the switching is carried out in sequence, the rotor will rotate with a stepped motion. However, as the switching speed increases, the rotor motion becomes continuous and thus the rotational speed can also be controlled by the switching process. The rotational direction will be reversed if the switching is carried out in the reverse sequence.

According to an article 'The application of electricity in warships' carried in an issue of *JIEE*,[1] this type of stepping motor was actually used to remote-control the direction indicator of torpedo tubes and guns in British warships. As illustrated in Fig. 1.9, a mechanical rotary switch was used for switching the exciting current. One revolution of the handle produces six stepping pulses causing 90° of rotor motion. The rotor motion in steps of 15° was geared down to attain the positional accuracy required.

As shown in Fig. 1.10, in place of the mechanical rotary switches, transistors or MOSFETs have recently been used for switching motor currents, and the switching signals are generated by a microprocessor. As the currents flow only downwards in this circuit when the power devices are closed, this type of

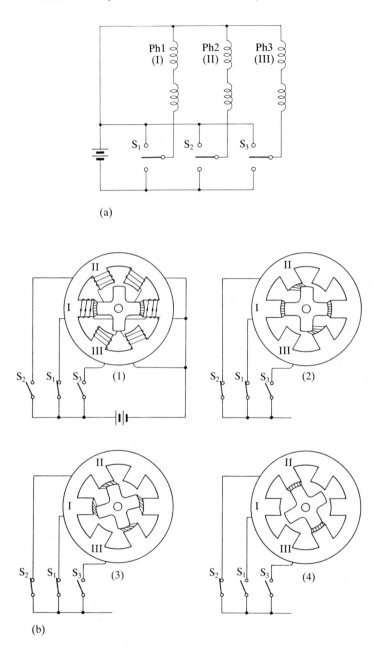

(a)

(b)

Fig. 1.8. (a) Connecting the stator coils in stator B for three-phase variable reluctance stepping motor operation with the universal manual switches in the MECHATRO LAB; (b) switching sequence and rotor positions.

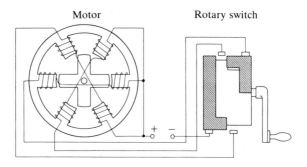

Fig. 1.9. Rotary switch used for driving a V R stepping motor in British warships in 1920s.

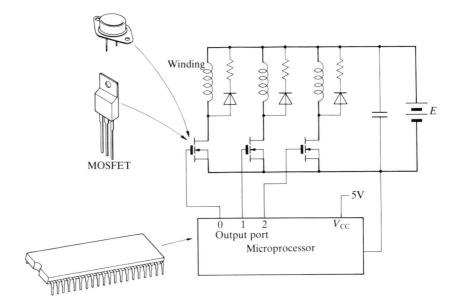

Fig. 1.10. Drive system of a stepping motor using a microprocessor for the unipolar scheme.

drive is referred to as 'unipolar' drive. It is possible to use the universal transistor circuit of Fig. 1.2 in unipolar fashion, using three transistors Tr_2, Tr_4 and Tr_6, and connecting power supply to the windings as illustrated in Fig. 1.11. A simple software operation for this motor in the sequence known as one-phase-on mode is given in the flowchart of Fig. 1.12. Here, Phase 1 is excited first; then it is de-energized while Phase 2 is excited, causing the rotor

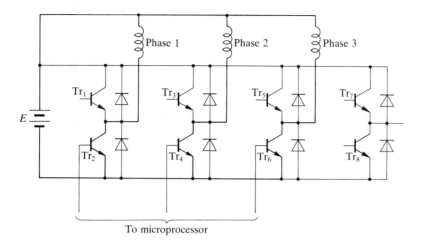

Fig. 1.11. Using the MECHATRO LAB transistor switches for the drive of a stepping motor in the unipolar scheme.

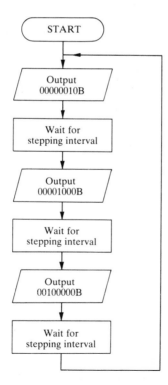

Fig. 1.12. Flowchart for driving a three-phase stepping motor in the one-phase-on mode at a constant stepping rate.

to travel 30° in a counter-clockwise direction; finally the current is commutated to Phase 3 to rotate the motor through a further 30°.

Table 1.1 gives an illustrative program listing when an external controller using a Z80 is to be employed for this experiment on the system of Fig. 1.2.

Table 1.1. Program listing for driving a three-phase stepping motor at a constant stepping rate. In this example the output port address is 0FDH, and the starting address is 8400H. Stepping rate is adjustable by changing the data loaded in register B or D in the TIME SUBROUTINE. (Applicable to Z80 and 8085.)

```
                              ORG     8400H

                         ;**** MAIN ROUTINE ****

        00FD                  DRIVE   EQU    0FDH

        8400   3E 02   LOOP:   LD     A,00000010B
        8402   D3 FD           OUT    (DRIVE),A
        8404   CD 8418         CALL   TIME
        8407   3E 08           LD     A,00001000B
        8409   D3 FD           OUT    (DRIVE),A
        840B   CD 8418         CALL   TIME
        840E   3E 20           LD     A,00100000B
        8410   D3 FD           OUT    (DRIVE),A
        8412   CD 8418         CALL   TIME
        8415   C3 8400         JP     LOOP

                         ;**** TIME SUBROUTINE ****

        8418   06 64   TIME:   LD     B,100
        841A   0E 64   LOOPA:  LD     C,100
        841C   0D      LOOPB:  DEC    C
        841D   C2 841C         JP     NZ,LOOPB
        8420   05              DEC    B
        8421   C2 841A         JP     NZ,LOOPA
        8424   C9              RET

                              END
```

1.3 Microprocessor-controlled power electronics drive of alternating-current motors

Motors that are run on alternating current are known as a.c. motors. There are two basic categories of a.c. motors; commutator motors and rotating-field motors. The a.c. commutator motor, which is also known as the universal motor as it can run on a d.c. supply too, is seldom used as a rotary machine to be controlled by electronic means. Hence in this book, only rotating-field motors are dealt with in association with electronic controls and drives.

1.3.1 *Functions of stators: devices to generate a rotating field*

The major components of a motor are the stator and the rotor. The stator core is formed with laminations of insulated silicon-steel plates, with windings installed in their slots. When alternating current flows in the windings, a rotating magnetic field is created that causes the rotor to revolve. We shall first examine how a rotating magnetic field can be created by the combination of the d.c. power supply and the manual switches or transistor power circuit in the MECHATRO LAB system shown in Fig. 1.2.

The simplest arrangement of so-called three-phase windings and their connections is illustrated in Fig. 1.13. In this example, one phase winding consists of only one coil, and the three coils are installed in six slots. The coil terminals may be either star-connected or delta-connected, as shown in Figs 1.13(b) and (c), respectively. The following explanation will be for the star connection, which is also known as the Y connection.

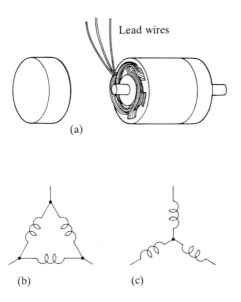

Fig. 1.13. (a) Simplest arrangement of the three-phase windings and (b) the ways of connecting coil terminals (delta (Δ) and star (Y) connection).

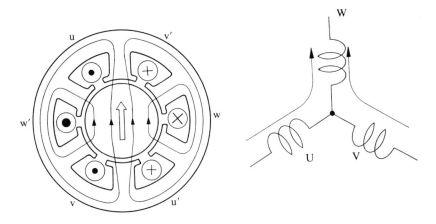

Fig. 1.14. Current distribution and flux path in a Y-connected, two-pole three-phase motor with terminals U and V connected to a positive power supply and W to the GND.

Figure 1.14 shows how the currents are distributed in the windings and how a magnetic field is formed in the motor when the U and V terminals are connected to the positive terminal of the battery and the W to the negative terminal. It is seen that two magnetic poles, the North and South poles, are created in this winding arrangement. This field distribution can be represented by a vector, illustrated in the centre of the rotor (as in Fig. 1.15). The vector has a magnitude and a direction.

Figure 1.15 illustrates the relationship between six different switching states and the magnetic field vector. There are other possible switching states but these six can be used in sequence to rotate the field; when switching is carried out downwards, the field will rotate clockwise (CW); with switching sequence upwards, the field will rotate counter-clockwise (CCW). In either sequence, one switching action makes the field travel through 60°, and hence six sequential operations make one full rotation. When a permanent-magnet rotor having two poles, as shown in Fig. 1.16, is placed in the rotating field, the rotor can be rotated by manual operation of the switches.

When the stator B in the MECHATRO LAB is used with the six coils connected in the two-pole scheme, a similar thing will take place as shown in the right column in Fig. 1.15. A 60° motion will be seen, if the rotor (4), which is the semihard steel rotor, is put in the stator and the switching action is carried out. As will be explained later, the number of the magnetic poles in a semihard steel rotor becomes the same as that in the stator.

In many a.c. motors, the stator core has more slots and teeth, and winding connections are more complex. The stator used in MECHATRO LAB has

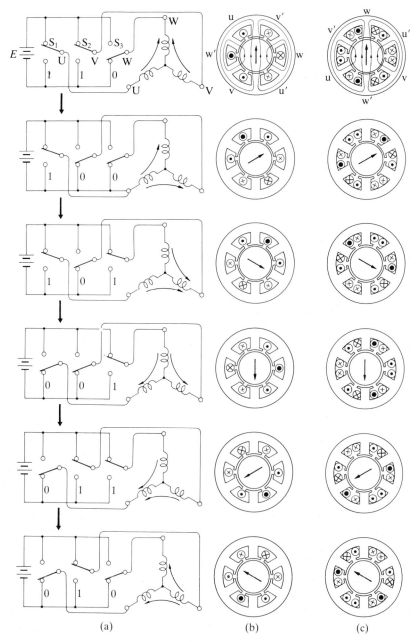

Fig. 1.15. Relation between six different switching states and magnetic field vector: (a) switching states; (b) for the stator in Fig. 1.13; (c) for stator B used in MECHATRO LAB.

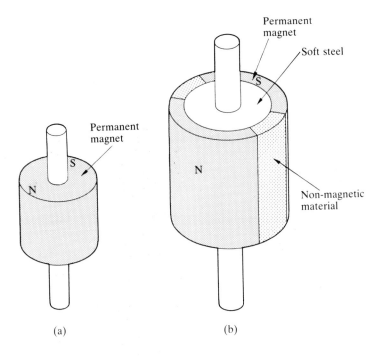

Fig. 1.16. Simple permanent-magnet rotors having two magnetic poles: (a) small rotor; (b) medium and large rotor.

four- and eight-pole windings on its 24-slot core. This example is very useful for understanding how practical windings are formed. The eight-pole arrangement is a simple extension of the two-pole winding just explained. As shown in Fig. 1.17(a), one phase winding consists of eight coils, and a typical arrangement for the three phases are shown in Fig. 1.17(b). It is seen in Fig. 1.18 that in an eight-pole motor the magnetic field travels through 15° with one operation of switch and therefore 24 sequential operations or four cycles make one full rotation of the magnetic field.

The 'goodness' of winding of an a.c. or brushless d.c. motor is often quoted by the number of slots per pole per phase. In the first example, having six slots for two-pole three-phase windings, the index is 1, since $6/2/3 = 1$. The index of the second case is also 1, since $24/8/3 = 1$. It is known that an index greater than 1 provides a better field distribution that produces less fluctuation in the travel of the field. This may be illustrated using the four-pole winding installed in the same core.

As the number of slots per pole per phase is now $24/6/4 = 2$, two coils can be accommodated in one pole in one phase, as shown in Fig. 1.19(a) and (b);

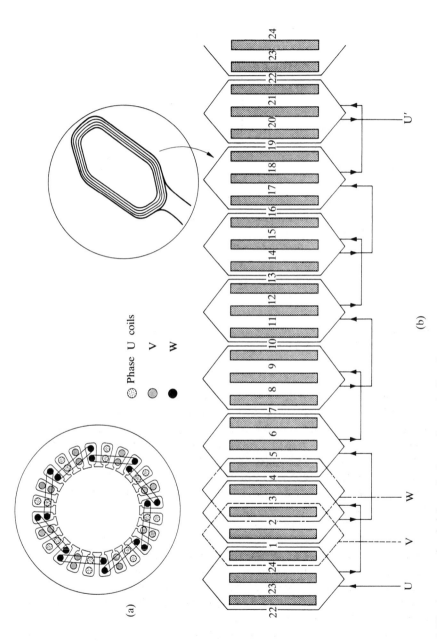

Fig. 1.17. An example of eight-pole three-phase windings in a 24-slot core: (a) cut-away views of phases U and V and the top view of phase W; (b) coil connections.

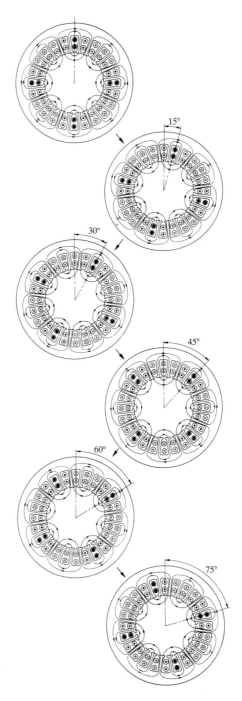

Fig. 1.18. Travel of the magnetic field in the switching operation of an eight-pole stator.

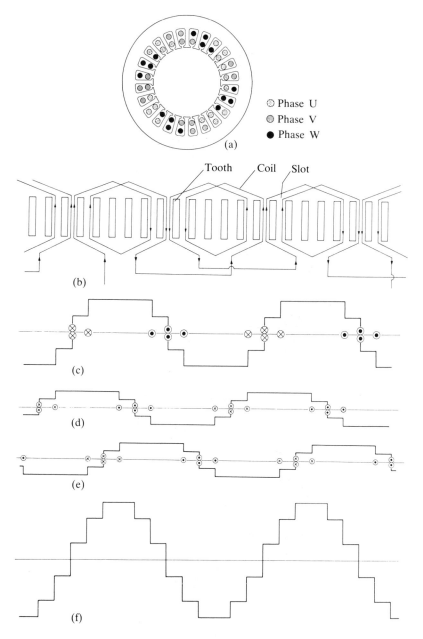

⊘ Phase U
◉ Phase V
● Phase W

Fig. 1.19. A typical four-pole winding in a 24-slot core, and the flux pattern: (a) cross-sectional view; (b) coil connections of phase U; (c) magnetomotive force distribution produced by a maximum current flowing in phase U; (d) magnetomotive force by phase V, in which a negative half-current flows; (e) magnetomotive force by phase W, in which a positive half-current flows; (f) the overall magnetomotive force distribution.

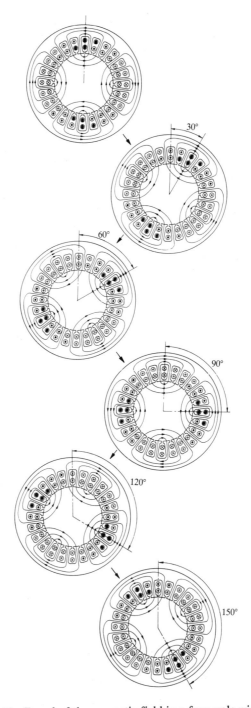

Fig. 1.20. Travel of the magnetic field in a four-pole winding.

two coils are arranged shifted by one slot-pitch to each other. Let us see how the magnetomotive force is created in the air-gap of the motor when currents flow as indicated in Fig. 1.19(b); a certain current flows in phase in the positive direction, and in the other two phases its half-current flows in the negative direction.

The waveform of the magnetomotive force created by the current in the winding of phase U is illustrated in Fig. 1.19(c), the secondary effect of slots on the field pattern being ignored. Figures 1.19(d) and (e) are the magneto-motive force distributions produced by the other two phases, and the addition of the magnetomotive forces of the three phases is shown in Fig. 1.19(f). Thus, the resultant flux distribution is near to a sinusoidal pattern.

As will be shown later, when the winding's spatial distribution is sinusoidal and when currents also sinusoidally varying are supplied to the windings, the magnetic field travels keeping a fixed pattern. Even when the motor is driven by the switching operation of transistors, the rotor motion is smoother with such a winding than in the case with an index of 1. It is seen in Fig. 1.20 that in a four-pole motor the field travels through 30° with each switching action.

1.3.2 *Power electronics drive of alternating-current motors*

When commutation of currents implemented by the mechanical switches shown in Fig. 1.15 is replaced by the combined function of transistorized

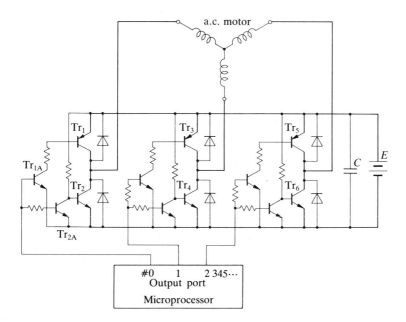

Fig. 1.21. A simple inverter using bipolar transistors and a microprocessor.

power electronics circuit and a microcomputer, we have one type of power electronic inverter for driving an a.c. motor. Details of inverter technology will be dealt with in Chapter 7; here, a simple inverter system is illustrated in Fig. 1.21. In contrast to the power electronics circuit of MECHATRO LAB, both NPN and PNP transistors are used in the so-called 'collector follower' scheme.

In this system, the microprocessor is used to generate the switching codes to operate the six transistors. For example, when the #0 bit is on the H level, Tr_{1A} is turned ON, and this causes Tr_1 to turn ON also. At this time, on the other hand, Tr_{2A} is ON and this makes Tr_2 OFF. If the output signal changes to the L level, the states of these transistors are all reversed. Similarly, the #1 and #2 bits of the output port drive the other two phases, and thus three bits are used for controlling the six transistors in this circuit. The first switching state in Fig. 1.15 is 110, the next is 100, and this is followed by 101, 001, 011 and 010. This sequence is repeated cyclically.

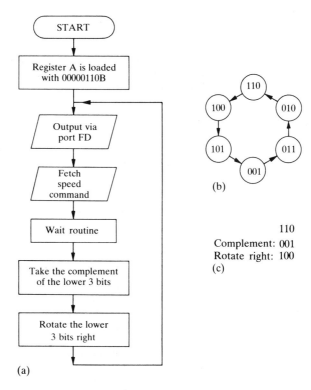

(a)

(b)

110
Complement: 001
Rotate right: 100
(c)

Fig. 1.22. Program flowchart of the inverter drive of an a.c. motor. (a) Flowchart for generating switching of signals; (b) switching code cycle; (c) how to progress one step.

The flowchart shown in Fig. 1.22(a) is a simple example for driving the motor in one direction at a variable speed following this sequence. The switching state codes are cyclically generated in the sequence shown in Fig. 1.22(b) on the principle explained in Fig. 1.22(c). A program listing based on this idea is given in Table 1.2. Since no protection measures are incorporated in the hardware and software in this example, it is recom-

Table 1.2. Program listing for a simple experiment on the inverter drive of an a.c. motor. The three-bit rotation is implemented in the ROTATE SUBROUTINE, and the frequency is adjustable by the data given from port whose location is 0FCH. (Applicable to Z80 only.)

```
                      ORG    8450H

              ; ****  MAIN ROUTINE ****

8450   0E FC          LD    C,0FCH    ;Define input port address
8452   3E 06          LD    A,00000110B ;Initial switching code
8454   A7             AND   A         ;Clear carry flag

8455   D3 FD   LOOP:  OUT   (0FDH),A  ;Output switching code
8457   ED 58          IN    E,(C)     ;Fetch speed data
8459   CD 8462        CALL  TIME      ;Spend a time
845C   CD 846D        CALL  ROTATE    ;Rotate lower 3 bits
845F   C3 8455        JP    LOOP      ;Go to LOOP

              ; ****  TIME SUBROUTINE ****

8462   16 64   TIME:  LD    D,100     ;Load D with a number (100)
8464   15      LOOPA: DEC   D         ;Decrement D till
8465   C2 8464        JP    NZ,LOOPA  ;D=0
8468   1D             DEC   E         ;Decrement E
8469   C2 8462        JP    NZ,TIME   ;If E≠0, then go to TIME
846C   C9             RET             ;Return to main routine

              ; ****  ROTATE SUBROUTINE ****

846D   2F      ROTATE: CPL            ;Invert (take complement of) A
846E   E6 07          AND   7         ;Clear higher 5 bits
8470   47             LD    B,A       ;Make copy of A in B
8471   E6 01          AND   1         ;Take LSB in A
8473   07             RLCA            ;Rotate left
8474   07             RLCA            ;Rotate left
8475   07             RLCA            ;Rotate left
8476   80             ADD   A,B       ;Add B to A
8477   0F             RRCA            ;Rotate right
8478   E6 07          AND   7         ;Clear higher 5 bits
847A   C9             RET

                      END
```

mended that an experiment should be carried out using a low-voltage power supply with a current limit at a low level.

1.4 Characteristics of alternating-current motors

The four major types of electrical motor are the conventional d.c. motor, the brushless d.c. motor, the stepping motor, and the a.c. motor. The first two types are dealt with in detail in Reference [2], and the stepping motor in Reference [1]. Some fundamental characteristics of a.c. motors are discussed in this section.

1.4.1 *Three major factors in forming a rotating field*

The following deals with three major considerations associated with rotating fields and windings.

(1) *Number of poles.* The number of magnetic poles that appear in a field pattern is referred to as the 'number of poles', and is an important factor that determines the motor's speed in conjunction with the power supply frequency. The number of poles is determined by the winding arrangements and connections. The most popular number of poles is four.

(2) *Number of phases.* Most motors are equipped with three-phase or two-phase windings according to the type of the power supply used. Stator A of MECHATRO LAB has a set of three-phase four-pole windings and another set of three-phase eight-pole windings.

(3) *Power supply.* Conventionally, most three-phase power motors are supplied from three-phase 50/60 Hz supplies that are generated by a power station and transmitted via lines and transformers. In this traditional drive, the currents vary sinusoidally with time, and if the spatial distribution of the windings is also sinusoidal, the magnetic field travels at a constant speed with a constant magnitude, as shown in Fig. 1.23. Hence, the rotor motion produces no prominent cogging except for some owing to the secondary effect between teeth on both rotor and stator.

Many small a.c. motors are designed to be driven by a single-phase supply. In the condenser-run motor, one type of popular single-phase motor, the single-phase current is converted into two-phase or three-phase current using the circuits shown in Fig. 1.24. It is, however, very inefficient to drive a single-phase a.c. motor over a wide speed range by this method. Many a.c. power motors and servomotors are now driven by electronic inverters to cover a wide speed and torque range for convenience of use and for energy-saving purposes.

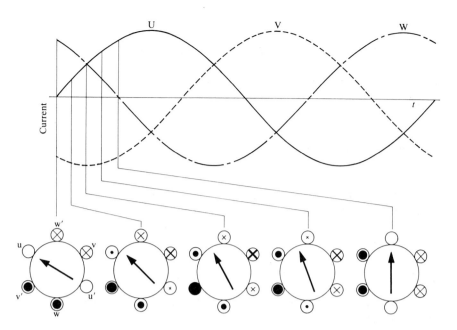

Fig. 1.23. When three-phase sine-wave currents are supplied to the windings, the magnetic field revolves at a constant speed with a constant magnitude, which makes the rotor move with little cogging. It should be noted that, in operation at normal frequencies, voltage is more important than the current in determining flux density.

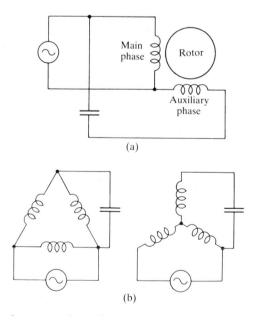

Fig. 1.24. Producing two or three-phase current from a single-phase source, using a capacitor for the condenser-run motor: (a) two-phase scheme; (b) three-phase schemes.

1.4.2 *Synchronous speed*

The rotational speed of the magnetic field in a motor is referred to as the synchronous speed. The synchronous speed N_0 does not depend on the number of phases, but is a function of the number of poles and the frequency given by

$$N_0 = 2f/p \text{ (r.p.s.)} \quad \text{or} \quad 120f/p \text{ (r.p.m.)} \tag{1.1}$$

where f = power supply frequency (Hz)
    p = number of poles
   r.p.s. = revolutions per second
   r.p.m. = revolutions per minute.

1.4.3 *Various constructions of rotors*

Classification of rotating-field type a.c. motors can be made on the basis of the rotor's basic structure. A summary of classification was shown in Fig. 1.6(b). Alternating-current motors are classified, depending on the rotor construction, largely into two groups: asynchronous and synchronous motors. In a synchronous motor the rotor speed is the same as the synchronous speed, while in an asynchronous motor the rotor speed is lower than the synchronous speed and depends on the load that the rotor carries.

Both asynchronous and sychronous motors can be sub-divided according to the material and structure employed in the rotor as follows:

(1) Asynchronous motors
    (a) squirrel-cage induction motor
    (b) wound induction motor
    (c) eddy-current motor
(2) Synchronous motors
    (a) wound-field synchronous motor
    (b) reluctance motor
    (c) permanent magnet motor
    (d) hysteresis motor

The construction of these practical rotors is well represented by the model rotors in Fig. 1.6(a), except for the wound induction motor and the wound-field synchronous motor. Figure 1.25(a) shows an experimental rotor that reasonably represents the practical rotor structure of these machines. It has a set of three-phase windings and three sliprings. When the machine is operated as an induction motor, a set of rheostats is connected to the brushes contacting the sliprings to form a circuit as shown in Fig. 1.25(b). It is known that the higher the resistance in the rheostats, the higher is the starting torque, and, conversely, the lower the resistance, the more efficient is the machine in the higher speed range. When the machine is used as a synchronous motor, a d.c. current must be fed from two terminals as depicted in Fig. 1.25(c).

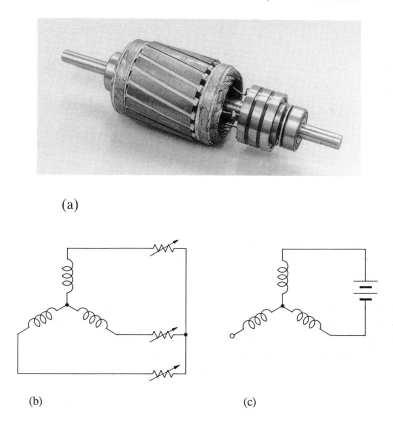

(a)

(b)   (c)

Fig. 1.25. (a) Experimental rotor used either as a wound induction motor or as a wound-field synchronous motor. Rotor winding connection for (b) induction motor and (c) synchronous motor operation.

Since the wound-field synchronous motor and the wound induction motor are classic motors, little interest has been paid to the electronic control of these machines. Therefore, a proper rotor for these machines is not included in MECHATRO LAB. Instead, rotor No. 6, having only a short-circuited four-pole winding, has been provided to be used to confirm that it can run as an induction motor.

1.4.4 *Tension of the lines of magnetic intensity generates the torque*

When explaining how to run the stepping motor by switching currents, using Fig. 1.8, the idea of tension in magnetic flux was used. Let us survey this theory briefly, using Fig. 1.26. When magnetic flux enters or leaves the

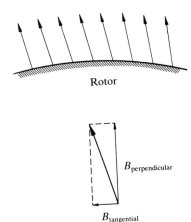

Fig. 1.26. When magnetic flux leaves the rotor surface at an angle with the surface that deviates from a right angle, a torque acts on the rotor. The tangential force per unit area on the rotor is $B_{tangential} B_{perpendicular}/\mu_0$.

rotor surface at an angle deviating from a right angle, one can say that a tangential force is produced on the surface of the rotor. Quantitatively, the force per unit surface is

$$F_s = B_{tangential} B_{perpendicular}/\mu_0 \qquad (1.2)$$

where μ_0 = the permeability of vacuum.

Qualitatively, one can consider that a strong tension is produced along the magnetic line of intensity, such as occurs on a string, and that its tangential component is utilized to produce a torque. However, as this tension is known to be $\frac{1}{2}B^2/\mu_0$, so that its tangential component is not equivalent to that given by eqn (1.2), we must be careful when developing this theory quantitatively. We shall not discuss this problem here, but the qualitative explanation is very convenient when we deal with both synchronous and induction motors.

Figure 1.27(a) shows the cross-section of a conductor placed in a homogeneous magnetic field. It is assumed that a current is flowing in the conductor at right angles to the plane of the paper and away from the reader. The magnetic field distribution created by the conductor current is concentric, as shown by solid lines. The farther the position from the conductor, the weaker the field intensity becomes. Figure 1.27(b) shows the field pattern as the result of vectorial addition of the homogeneous field. The magnetic lines of force show steep curvature at the periphery of the conductor. The curvature of the lines of force manifests itself as a tendency to straighten themselves owing to the magnetic tension. Thus, one can interpret this action as causing the conductor to be pushed towards the left.

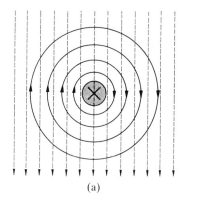 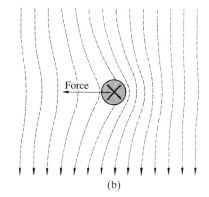

(a)   (b)

Fig. 1.27. A conductor placed in a homogeneous magnetic field. (a) When a current passes through the conductor, a concentric magnetic field is generated by this current. (b) The vectorial addition of these two fields shows curvature in the lines of intensity around the conductor.

Next we must consider the source and cause of the conductor current. This current may be supplied from a battery, but it can be caused by the movement of the magnetic field. For example, when the magnet producing the homogeneous field is travelling towards the left, as shown in Fig. 1.28, there is relative motion between the conductor and the field. Hence, it can be considered that the conductor is travelling towards the right in the magnetic field. As given by Fleming's right-hand rule, an electromotive force is generated in the conductor so as to cause the current to flow in the direction away from the reader. Thus, the movement of a magnetic field can cause a conductor placed in it to travel in the same direction.

1.4.5 Squirrel-cage induction motor

Of various types of asynchronous motors, the most important is the squirrel-cage induction motor. The most notable feature in the rotor structure is the conductor assembly, which is similar to the shape of a cage used for keeping squirrels or mice, as shown in Fig. 1.29(a). This assembly, consisting of a number of conductor bars and two endrings, is covered with laminated core of silicon steel, as shown in Fig. 1.29(b). The bars and endrings form current paths and the laminated core serves as the flux path. Figure 1.29(c) shows a typical construction of a squirrel-cage induction motor.

Let us assume that the rotor, placed in a rotating magnetic field, is initially at rest. As each bar is connected to each other bar via the endrings, current can flow in either direction in each bar, depending on the polarity induced.

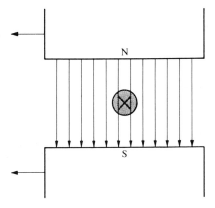

Fig. 1.28. When the homogeneous field moves towards the left, this causes a current to flow in the direction away from the reader in the conductor.

As already seen, owing to the relative motion between the revolving magnetic field and the conductor bars, a torque is applied to each bar, the whole rotor being caused to rotate in the same direction as the revolving field.

Even after the rotor has started to move, as long as the rotor speed is lower than the synchronous speed, a torque is produced that accelerates the rotor until the torque is balanced by the load the rotor carries.

The torque-versus-speed characteristics of a squirrel-cage induction motor driven by a constant-voltage constant-frequency supply are as shown in Fig. 1.30(a). Generally, the lower the resistance of the rotor conductors, the lower the starting torque but the higher the efficiency in higher speed ranges.

When this motor is run using a variable-frequency inverter with the voltage-to-frequency ratio kept constant, the torque-versus-speed characteristics will be as shown Fig. 1.30(b). It is seen that by starting the motor at a low frequency, a high starting torque is available with low current consumed, and that by increasing the frequency as the motor accelerates the machine can be operated at optimum conditions.

1.4.6 *Eddy-current motors*

There is a type of induction motor that uses, instead of a squirrel-cage rotor, a rotor whose main material is mild steel, as shown in Fig. 1.6(a)(1). This motor is known as the eddy-current or solid-steel induction motor. Unlike the case of squirrel-cage rotor, the core serves as both the current path and the flux path. The torque-versus-speed curves of an eddy-current motor display drooping characteristics, as shown in Fig. 1.31. A typical real eddy-current motor is an outer-rotor motor as shown in Fig. 1.32.

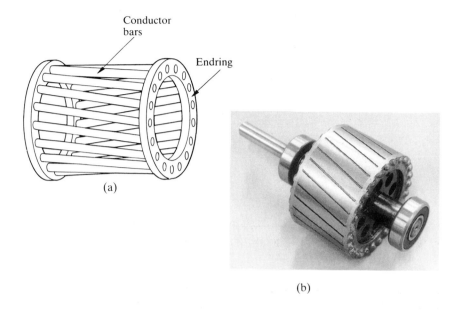

(a)

(b)

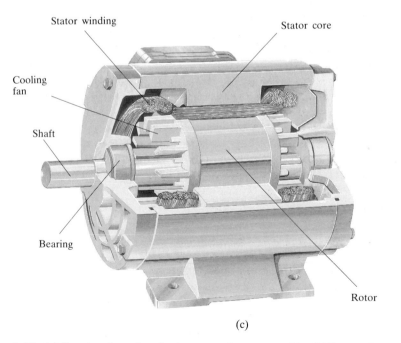

(c)

Fig. 1.29. (a) Construction of squirrel-cage conductor assembly. (b) Rotor of a small motor. (c) A cutaway view of a medium-size induction motor.

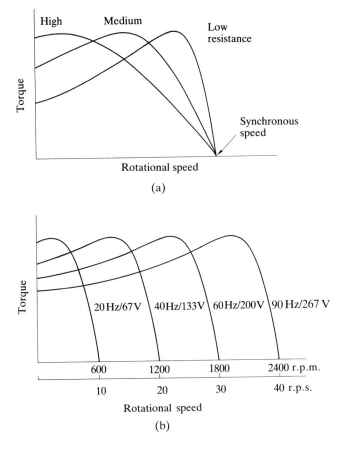

Fig. 1.30. Torque-versus-speed characteristics of a squirrel-cage induction motor. (a) Effect of the resistance of the squirrel-cage assembly under constant frequency operation. (b) Variable-frequency variable-voltage inverter drive with a constant frequency-to-voltage ratio.

Advantages of the solid-steel motor are that no special material is required for the rotor and that it is free of the cogging that often occurs in a squirrel-cage motor owing to the mutual effects of teeth on both the stator and rotor.

1.4.7 *Reluctance motor*

The reluctance synchronous motor, also known as reaction motor, has a so-called 'salient poled' rotor in principle. As seen in Fig. 1.6(b)(3), the rotor's cross-section consists of convex and concave portions. The convex portions

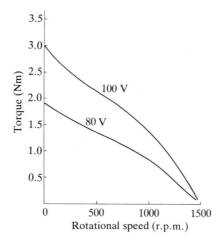

Fig. 1.31. An example of torque-versus-speed characteristics of an eddy-current motor driven on a 50 Hz three-phase supply.

are called the salient poles, and as a general rule, the number of salient poles must be equivalent to the number of magnetic poles produced by the stator windings and currents.

Figure 1.33 illustrates the fundamental principle of the two-pole reluctance motor. Suppose that the rotor is placed in a magnetic field created by a two-pole permanent magnet, and that the rotor material is a mild steel having a high permeability. Referring to the figure, let us assume that the rotor's convex axis BO forms a certain angle with respect to the stator's magnetic pole axis AO. Now, the magnetic flux produced by the permanent magnet will pass through the mild steel in the rotor. Since the two axes form an angle, the magnetic field in the air-gaps will be inclined or curving. Owing to this inclination, a torque is produced in such a direction that the stator axis AO and rotor axis BO will be brought into alignment. Therefore, when the outer permanent magnet is caused to revolve, the rotor will revolve so as to follow it. By replacing the outer permanent magnet by a set of windings that can produce a rotating field, a practical reluctance motor can be built.

However, there is a problem here: the motor will have no self-starting capability if its construction is based on the basic principle only. That is, even when the magnetic field starts to revolve at switching-on, the rotor will not be capable of rapidly beginning rotation, owing to its inertia. The rotor will only vibrate with a clattering noise and generate heat. Ordinarily, to provide the motor with self-starting capability, a squirrel-cage rotor having salient poles or a construction as illustrated in Fig. 1.34 is adopted. Figure 1.35 shows the torque-versus-speed characteristics of a reluctance motor.

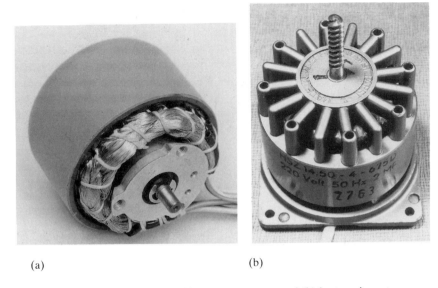

(a)   (b)

Fig. 1.32. (a) Outer-rotor eddy-current motor, and (b) hysteresis motor.

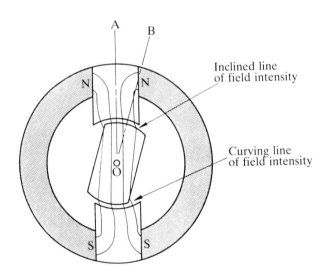

Fig. 1.33. Principle of the reluctance motor

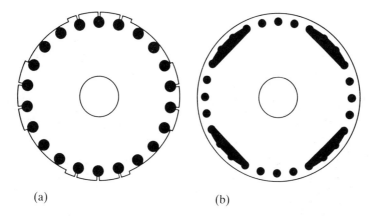

Fig. 1.34. Cross-sections of reluctance motor: (a) has concave and convex portions and (b) is cylindrical.

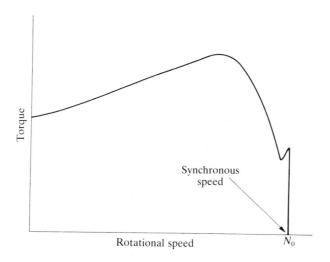

Fig. 1.35. Torque-versus-speed characteristics of a reluctance motor.

1.4.8 *Permanent-magnet motor*

As its name implies, the permanent-magnet motor is a motor that employs a permanent magnet in its rotor. Figure 1.36 shows a situation in which a cylindrical permanent-magnet rotor is placed in a two-pole magnetic field. From the drawing it may easily be understood that a torque will work on the magnet rotor if the stator field is rotated by employing windings and polyphase currents. When this situation is compared with that of the reluctance motor described earlier, the following will hold:

1. When a permanent magnet is used as the rotor, the shape may be cylindrical. (Of course the rotor may employ dovetail-shaped magnets, like the rotor in Fig. 1.6(b)(5) or that of Fig. 1.16. Some other shapes of magnet can also be used, as shown in Fig. 8.27.)
2. When mild steel is used as the rotor material, its shape must have saliency in order to be operated as a synchronous motor.

Like the reluctance motor, the permanent magnet motor in its pure form has no self-starting capability when driven as a plain a.c. motor. For this reason, in practice, a rotor having the construction shown in Fig. 1.37, which is a four-poled structure, is employed to aid self-starting.

A small permanent-magnet motor offers much higher efficiency than a reluctance motor or a hysteresis motor of the same size. It should also be

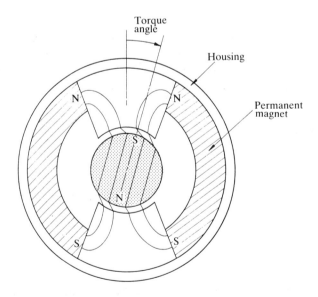

Fig. 1.36. Permanent-magnet rotor placed in a two-pole stator.

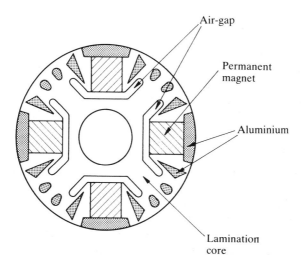

Fig. 1.37. Rotors using combined principles of squirrel-cage induction, reluctance, and permanent magnet rotor (after Reference [3]).

noted that a permanent-magnet motor is often used in a brushless-drive scheme.

1.4.9 *Synchroduction rotor*

A unique permanent-magnet rotor proposed by D. McGee[4] is shown in Fig. 1.38. This is a combination of three of the different principles so far discussed. The permanent magnets mounted at the rotor ends provide the main flux. The salient poles stretched from the shaft and the bridges in the sleeve form a reluctance structure, and the soft iron core and conductor material used to fill in the cavity (or the gap between the bridges and salient poles) can be the paths for the induced eddy-currents, to yield an induction-motor torque.

When an electromagnet is used instead of a permanent magnet, the field flux can be controlled to cover a wide range of speed when driven in the brushless d.c. motor scheme. A merit of this structure is that the permanent magnet and the electromagnet can both be stationary, when they are mounted on the stator or the end-plate.

1.4.10 *Hysteresis motor*

As the principle of this motor is very subtle but is dealt with all too briefly in most textbooks, a detailed explanation, a translation from part of one of

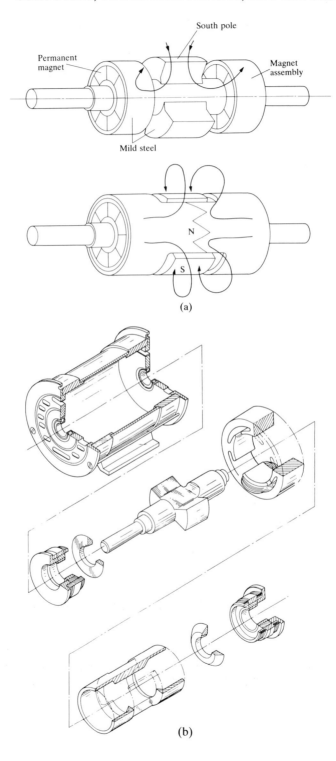

(a)

(b)

the author's Japanese books,[5] will be given here without much use of mathematics.

(1) *Properties of rotor material.*   A ring-shaped substance displaying magnetic hysteresis is used as the main material for the rotor (see Fig. 1.39(a)). Most hysteresis motors are normal inner-rotor type, a unique outer-rotor version as shown in Fig. 1.32(b) was once manufactured.

(a)

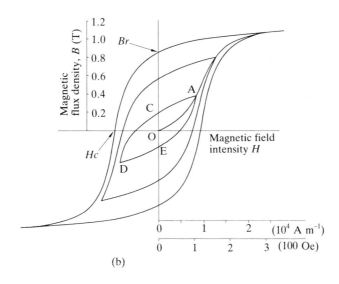

(b)

Fig. 1.39. (a) Hysteresis ring core of semi-hard steel; (b) magnetic hysteresis curves.

Fig. 1.38. A unique rotor construction proposed by D. McGee: (a) Magnetic sources are placed at the rotor ends, and salient poles of mild solid steel face the stator core; (b) exploded views.

First, it is necessary to make clear what kind of hysteresis property is utilized in this motor. Magnetic hysteresis comprises the flux-versus-field intensity characteristics that are represented by the curves in Fig. 1.39(b). Initially the B/H state is in the unmagnetized state or on the origin of the coordinates. When the magnetic field intensity H is applied and increased, the flux density B will increase along the so-called maiden curve and eventually reach point A.

As the field intensity is subsequently gradually decreased from this point, the B/H characteristics do not follow the original curve; instead, the flux density B will decrease along the curve AC. After the field intensity has been decreased to zero, it is next increased with the reverse polarity up to the point at which the intensity is opposite to but at the same value as at point A. When H is again decreased to zero from this point, denoted by D, and again increased with the previous polarity, the characteristics will not follow the previous curve, but instead will follow another curve DEA and return to point A. The closed loop that has been plotted as ACDEA is referred to as the hysteresis loop. A larger loop can be obtained by increasing the swing amplitude of the field intensity and following the same procedure as before. However, when magnetic saturation occurs, the hysteresis loop will not become any larger than a certain size.

The value of B at the point where the maximum hysteresis loop intersects the ordinate is referred to as magnetic remanence Br, while the value of H at which the curve intersects the abscissa is referred to as the coercive force or intensity Hc. For hysteresis motors, materials cast from alloys of iron, nickel, and aluminium, or other sorts of magnetic materials whose coercive intensity is in the range of 8–16×10^3 A m^{-1} are employed. Such values of coercive intensity are much lower than those of normal permanent magnets.

To obtain a large torque from a motor, it is desirable that the loop cover as large an area as possible and that its shape be close to an ellipse. As the loop gets larger, the shape will be distorted and depart further from elliptical.

(2) *What does hysteresis mean in torque production?* When we consider the rotor to be at rest, each point in the rotor is subjected to hysteresis due to the rotating magnetic field. Let us formulate the relation between B and H as a function of time. For the sake of simplicity, suppose the shape of the hysteresis loop to be elliptical, as shown in Fig. 1.40(b). If we assume that the field intensity at each point is varying in the form of a sine wave, as shown by the vertical solid curve, the corresponding variation of flux density B will be such as shown by the horizontal solid curve. In other words, the flux density also varies in the form of a sine curve.

To compare the H curve with the B curve, let us move the H-curve over to the B-curve by rotating it by 90°. This is illustrated by the chain-line horizontal curve. It is seen that there is a phase difference between B and H. Thus,

Hysteresis loop

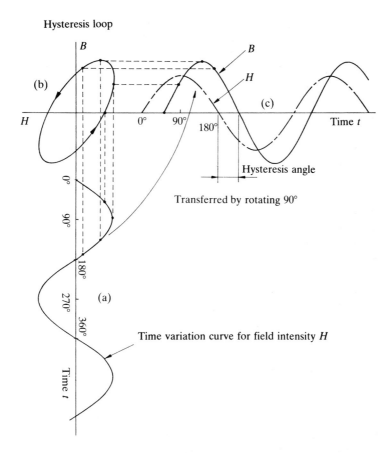

Fig. 1.40. A hysteresis loop represented by an ellipse, and time variation curves for flux density *B* and field intensity *H*.

hysteresis may be interpreted as the phenomenon in which *B* lags behind *H* in terms of phase. This is closely related to the torque production, as will be explained later.

(3) *What will the magnetic field distribution be inside the motor?* How can it be illustrated that a phase difference exists between *B* and *H* inside the rotor? The answer is shown Fig. 1.41. In practice, as shown in Fig.1.41(a), a ring of hysteresis material is employed and is supported by a non-magnetic hub to be fixed to the shaft. However, it is not easy accurately to illustrate the relation between field intensity distribution and flux density in a ring. Consequently, Fig. 1.41(b) shows a cylindrical rotor that is made of homo-

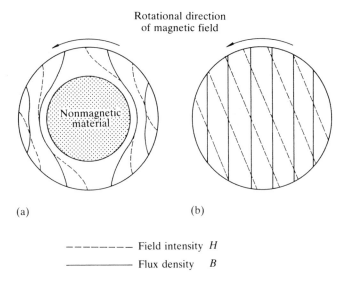

Fig. 1.41. Field intensity and flux distributions in semi-hard magnetic steel cores: (a) ring core; (b) cylindrical core.

geneous material. Even the shaft is assumed to be of the same hysteresis material. In this rotor, the distribution of B, as well as that of H, can be represented by straight parallel lines, and the group of lines of field intensity H and the corresponding group of flux density B form a certain angle equivalent to the phase difference angle between H and B. The patterns of H and B are both revolving, and the pattern of B is lagging behind that of the field intensity.

However, when such a cylindrical hysteresis material is used, a hysteresis occurs, known as rotating hysteresis, that is different from the ordinary alternating hysteresis illustrated in Fig. 1.40. In alternating hysteresis, the direction of the magnetic field is fixed, whereas the magnitude and polarity vary. In contrast, in rotational hysteresis, the magnitude of the field intensity is fixed, and the direction varies as the magnetic field rotates.

For a hysteresis motor, either alternating or rotating hysteresis is usable. To realize alternating hysteresis, however, the use of an extremely thin ring is required. Naturally, since the ring must be appreciably thick in a practical motor, it is assumed that the hysteresis that occurs is phenomenon intermediate between alternating and rotational hysteresis.

In the stator core, which is free from hysteresis, B and H are always in the same phase. If the rotor is assumed to be a lamination core of mild steel, the field intensity vector H produced by the current in the windings placed in a homogeneous air-gap displays a distribution as shown in Fig. 1.42(a). Since

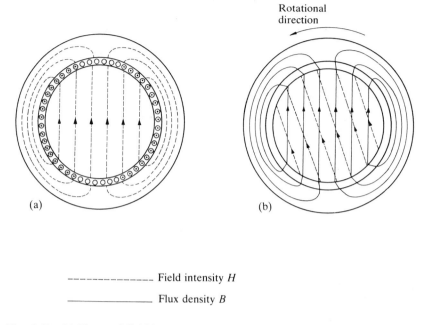

Field intensity H (dashed line)

Flux density B (solid line)

Fig. 1.42. (a) Flux and field intensity distribution when a soft-steel core is used for the rotor. (b) The case in which a hysteresis core is used in the rotor.

the whole core section is made of mild steel, the flow of the flux B is parallel to H. Again, when the rotor material displays a hysteresis phenomenon, there is a phase difference between B and H, while in the stator core placed outside the air-gap there is no phase difference. This situation is illustrated in Fig. 1.42(b).

As it is known that the flow of B is continuous at boundaries, it can be understood qualitatively that an inclination of field vector occurs in the air-gap. Careful calculations can be made by solving field equations, and the result tells us that the lines of intensity are not perpendicular to the rotor surface but are inclined a little in the gap. The direction of the torque generated owing to the inclination of the lines of intensity is the same as the rotational direction of the magnetic field pattern.

(4) *The hysteresis motor also has the property of an asynchronous motor.* The above explanation assumed the rotor to be at rest; that is, it described the situation for the starting torque. What will happen once rotation is initiated by this force? As long as the rotational speed is lower than the synchronous speed, the rotor material will still be subjected to hysteresis. But the speed of field variation will decrease. What is important in the preceding explanation is that a hysteresis takes place, but the frequency of hysteresis is not a problem. Hence, at speeds below synchronism, the same torque will be created as the starting torque. Thus, as already made clear, the hysteresis motor exerts a useful smooth torque from starting to synchronism. Therefore, if the load torque balances with the motor torque at a speed less than the synchronous speed, the motor will revolve stably at this speed.

When the motor reaches the synchronous speed, the motor torque will automatically be reduced to a level to balance with the load torque. Thus, the hysteresis motor has the properties of a synchronous motor. It should be noted that only the hysteresis motor achieves the properties of both the synchronous and the asynchronous motor via a single principle. In this respect, the hysteresis motor is quite different from the reluctance motor, in which the torque caused by salient poles is effective only at the synchronous speed and the rotor must be equipped with a squirrel-cage conductors to permit the motor to start and accelerate up to the synchronous speed.

(5) *Torque characteristics.* According to the above theory, the hysteresis motor will exhibit simple torque characteristics as illustrated in Fig. 1.43(a). Actually, however, the characteristics show a drooping tendency as depicted in Fig. 1.43(b). The reason for this is twofold:

1. The hysteresis ring is usually solid and conductive, which produces drooping characteristics.
2. Torque reduction is caused by the stator teeth. Because of the teeth in the stator core, ripples are produced in the flux on the rotor surface. This

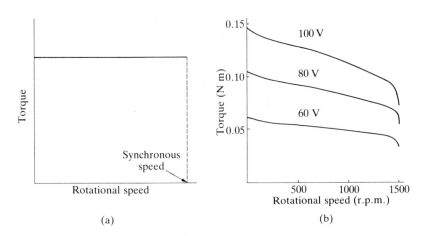

Fig. 1.43. Torque-versus-speed characteristics of a hysteresis motor: (a) ideal and (b) a real four-pole hysteresis motor driven at 50 Hz.

causes the torque to decrease in the speed range near the synchronous speed.

1.5 Microprocessor-controlled power electronics in driving a brushless direct-current motor

In the inverter drive of a squirrel-cage rotor, the switching signals are sent to the transistor circuit from an external signal generator. If one attempts to drive a permanent-magnet rotor using such an inverter, one will find the rotor will not start unless the frequency is extremely low. However, if a position sensor is mounted on the motor as illustrated in Fig. 1.44, and the position information is used in generating the switching signals, the motor can start and accelerate.

Figure 1.45 is the block diagram of a typical brushless d.c. motor. In operation of this system, the switching codes for the six transistors are determined by reference not only to the position information but also to the drive commands such as CW, CCW, braking, and so forth. The torque-versus-

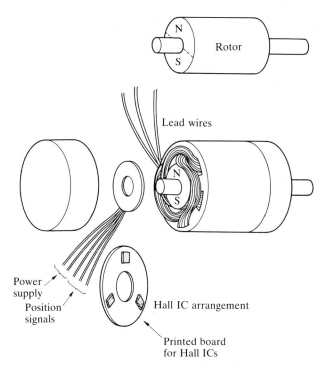

Fig. 1.44. Construction of a simple brushless d.c. motor.

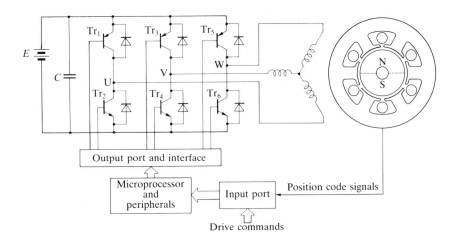

Fig. 1.45. Brushless d.c. drive system using a microprocessor.

speed characteristics obtained in this arrangement are normally very similar to those seen in a conventional d.c. motor having a commutator and brushes.

Details of the Hall elements and techniques of construction of brushless d.c. motor drives will be discussed in Chapter 8.

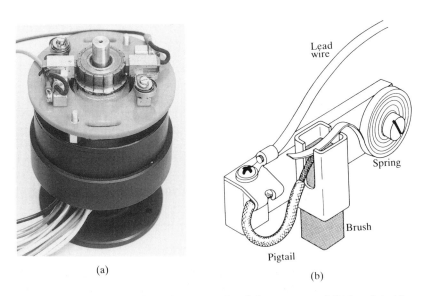

Fig. 1.46. (a) Construction of a conventional d.c. motor and (b) brush holder.

1.6 Microprocessor-controlled power electronics in the drive of a direct-current motor

Using the commutator rotor, brush holders, and stator B, one can construct a typical conventional d.c. motor as illustrated in Fig. 1.46(a). Figure 1.46(b) shows the construction of the brush holders that are necessary for fulfilling two functions: one is detecting the rotor's angular position or that of the coils to be supplied with current, and the other is switching the current from other coils to these in cooperation with the commutator.

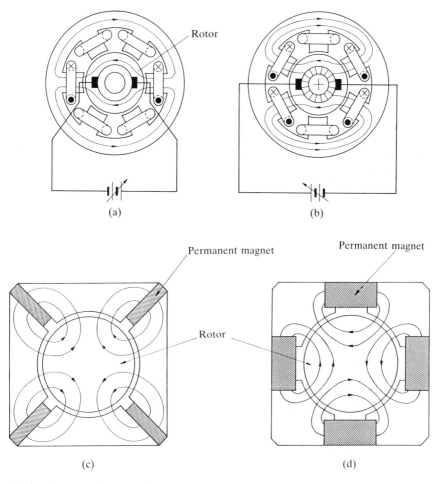

Fig. 1.47. (a) Using two coils of stator B, and (b) using four coils for providing field flux on stator B. (c) and (d) Cutaway diagrams of typical four-pole d.c. motors using permanent magnets.

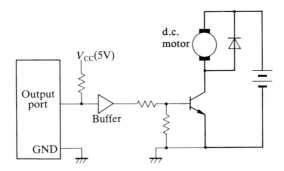

Fig. 1.48. A simple transistor circuit for a speed-control experiment in the pulse-width modulation drive of a d.c. motor.

As for the coil connections for creating the magnetic field, two or four coils of the six can be used, as shown in Fig. 1.47 (a) and (b). These figures show how to apply a d.c. power supply to the rotor. One may use the same supply or another to provide the stator coils with a current. Figure 1.47 (c) and (d) show cutaway views of typical four-pole d.c. motors using permanent magnets. The potential applied to the rotor can be controlled, in the pulse-width modulation (PWM) mode, by a simple circuit shown in Fig. 1.48. An equivalent electronic circuit can be produced using one transistor, for example Tr_2, in the power circuit in the MECHATRO LAB.

The transistor is operated in the high-frequency switching mode by the signal generated by the microprocessor, and the average potential applied to the motor is proportional to the pulse-width ratio defined by

$$\text{Pulse-width ratio} = \frac{T_{on}}{T_{on} + T_{off}} \tag{1.3}$$

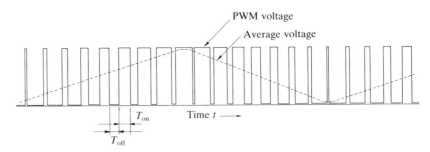

Fig. 1.49. Varying the speed of a d.c. motor by the pulse-width modulation signals created by a microprocessor.

where T_{on} = period for the transistor to be closed in a cycle
       T_{off} = period for the transistor to be open in a cycle.

As the no-load speed of a d.c. motor is proportional to the voltage across its terminals, the speed can be controlled by the adjustment of the pulse-width ratio.

Table 1.3. Program listing for generating PWM signal for variable-speed drive of a conventional d.c. motor. (Applicable to Z80 and 8085.)

```
                          ORG     8500H

                     ;****  MAIN ROUTINE ****

00FD                 DRIVE   EQU    0FDH      ;Declare output port
8500    01 0100      START:  LD     BC,0100H  ;Load B with 1 and
                                              ;load C with 0 (=256)
8503    3E 01        ON1:    LD     A,1
8505    D3 FD                OUT    (DRIVE),A ;Output ON signal
8507    50                   LD     D,B
8508    CD 8530              CALL   WIDTH     ;Generate pulse width
850B    3E 00        OFF1:   LD     A,0
850D    D3 FD                OUT    (DRIVE),A ;Output OFF signal
850F    51                   LD     D,C
8510    CD 8530              CALL   WIDTH
8513    04                   INC    B         ;Increment B
8514    0D                   DEC    C         ;Decrement C
8515    C2 8503              JP     NZ,ON1    ;If zero, then go to ON1,
                                              ;and else go next
8518    3E 01        ON2:    LD     A,1
851A    D3 FD                OUT    (DRIVE),A
851C    51                   LD     D,C
851D    CD 8530              CALL   WIDTH
8520    3E 00        OFF2:   LD     A,0
8522    D3 FD                OUT    (DRIVE),A
8524    50                   LD     D,B
8525    CD 8530              CALL   WIDTH
8528    04                   INC    B         ;Increment B
8529    0D                   DEC    C         ;Decrement C
852A    C2 8518              JP     NZ,ON2    ;If not zero, then go to ON2,
852D    C3 8500              JP     START     ;and else go to START

                     ;**** PULSE WIDTH SUBROUTINE ****

8530    15           WIDTH:  DEC    D         ;Decrement D
8531    00                   NOP              ;Adjust time by number of NOP
8532    00                   NOP              ;instruction
8533    00                   NOP
8534    00                   NOP
8535    C2 8530              JP     NZ, WIDTH ;If D=0, then go to WIDTH
8538    C9                   RET              ;Return to main routine

                          END
```

When the pulse-width ratio increases and decreases alternately, as illustrated in Fig. 1.49, the motor's speed will also show repetition of increase and decrease, though with a time lag. An example of the assembly-language program for a Z80 to generate a train of switching signal pulses for such a PWM drive is given in Table 1.3.

The theory of pulse-width modulation will be discussed in Chapter 4, and details on servo-amplifiers for both linear and PWM drives of conventional d.c. machines will be given in Chapter 5.

References

1. Kenjo, T. (1984). *Stepping motors and their microprocessor controls*. Oxford University Press, Oxford.
2. Kenjo, T. and Nagamori, S. (1985). *Permanent-magnet and brushless DC motors*. Oxford University Press, Oxford.
3. Binns, K. J., Molyneux, D.C., and Barnard, W. R. (1976). Some aspects of the development and design of a high performance permanent magnet synchronous motor. In *Small Electrical Machines*, IEE Conference Publication, No. 136, pp. 76–81.
4. McGee, D. (1987). Re-programming DC brushless motors. In *Proceedings of the 11th Motor-Con*. SATECH '87, pp. 58–65.
5. Kenjo, T. (1979). *Precision motors used in electronic equipment*. (In Japanese). Chapter 2. Sogo Electronics Publishing Company, Tokyo.

2 Solid-state devices as switch elements

In power electronics, solid-state devices are used mainly as switches, whereas they are used both as switches and as amplifiers in information electronics. Solid-state devices are often compared to mechanical switches because of their similarities, and it is believed that solid-state devices are superior to mechanical ones because they have very long lifetimes and can be switched at frequencies as high as several tens of kilohertz. However, solid-state devices are more delicate and more complicated in their properties and we must be careful in using them. In this chapter, the basic physical configuration and some properties of solid-state devices will be discussed, and followed by an explanation of some fundamental circuits for switching d.c. and a.c. current.

2.1 Impurity semiconductor materials and the P N junctions

Readers are expected to have some knowledge of semiconductor materials and their physics. However, the most basic and necessary parts of these subjects will first be surveyed.

Most solid-state devices are made of crystalline silicon. Silicon is an element belonging to the fourth column of the periodic table. This means that silicon atoms possess four electrons in their valence or outermost orbital, as shown in Fig. 2.1(a). The crystal structure of silicon is similar to that of diamond, as illustrated in Fig. 2.1(b). In this structure, each atom shares one of these valence electrons with each of its four neighbours. In such a covalent structure it is likely that each atom has eight valence electrons. It is hoped that readers recall that when atoms possess eight electrons in their valence shell they are very stable. Such a crystal structure is often illustrated by the two-dimensional representation given in Fig. 2.1(c).

Pure or intrinsic silicon is of little value as a semiconductor device because it has very low conductivity owing to the stable crystal structure. To be used in a diode or transistor, the semiconductor must possess impurities at small concentrations.

If, during the production of single crystal of silicon, some atoms of an element from the fifth column of the periodic table are inserted, these atoms will take up various positions throughout the crystal (see Fig. 2.2(a)). As each impurity atom has four immediate neighbouring silicon atoms, there are four valence electrons from these atoms available for forming covalent bonds. Hence, one of the electrons belonging to the impurity atom is not utilized in a

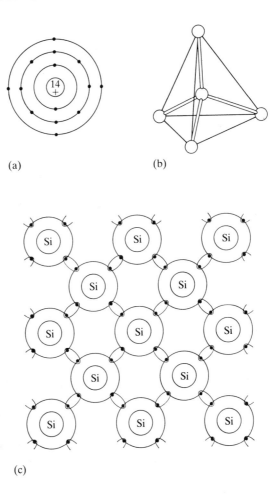

Fig. 2.1. (a) Four of the fourteen electrons in a silicon atom are in the valence or outermost orbit; (b) The crystal structure of silicon is similar to the very stable diamond structure of carbon. Each atom is surrounded by four immediate neighbours. (c) This arrangement is illustrated in two-dimensional presentation.

covalent bond. These surplus electrons are loosely bound to the parent atoms, and can be free to move about at room temperature. This type of doped semiconductor is called N-type, where N is derived from the negative charge of the surplus particle.

The other type of doped semiconductor is called a P-type material. This is formed by doping with atoms of an element from the third column of the periodic table during production of the crystal (see Fig. 2.2(b)). Since the tri-

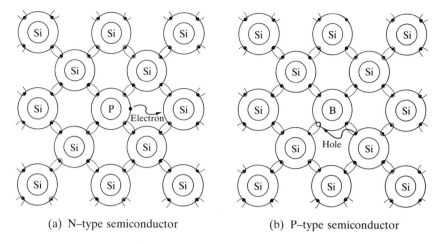

(a) N–type semiconductor   (b) P–type semiconductor

Fig. 2.2. Silicon crystal with impurity atoms: (a) When pentavalent atoms are inserted, their fifth valence electrons are loosely bound to their parent atoms and can become negative charge carriers. (b) If trivalent atoms are added, the vacancy, or deficit of a valent electron—known as a 'hole'—can freely move about owing to thermal energy at room temperature and can behave as positive charge carrier.

valent impurity atom is surrounded by four immediate silicon atoms, one electron is missing from a potential covalent bond. This deficit of an electron behaves as a positive charge carrier, and is known as a hole. Similarly to the excess electrons in the N-type impurity material, holes in a P-type material are loosely bound to their parent atoms and can be free to move in the material when an electrical field is applied at room temperature.

In most semiconductor devices, both N-type and P-type regions exist in a single crystal. The most frequently used dopant for P-type material is boron (B), while typical dopants for N-type materials are phosphorous (P) and arsenic (As). The transition zone from N-type to P-type is referred to as a PN junction. A PN junction has an important function, and solid-state devices have one or more PN junctions.

2.2 Types of solid-state devices

There are many different types of solid-state devices. The following are those that have long been used in power electronics equipment:

(1) diodes;
(2) Bipolar transistors;

(3) Power MOSFETs (metal oxide semiconductor field-effect transistors);
(4) The thyristor group, which is classified into:
   (a) thyristor (also known as reverse-blocking thyristor or SCR, standing for silicon controlled rectifier);
   (b) GTO thyristors (gate turn-off thyristors);
   (c) triacs (also known as a.c. thyristors).

The symbols for and basic characteristics of these devices (excepting the GTO and triac) are summarized in Table 2.1.

Table 2.1. Main solid state devices

Device	Symbol	Characteristics
Diode	Anode Cathode	The simplest device having rectifier characteristics Power-handling capability is the highest of all the solid-state devices (e.g. 4000 V/3000 A) There are special types of diodes such as Zener diodes, fast reverse-recovery diodes, etc.
Bipolar transistors	C B E Collector Base Emitter	Switching is implemented by the base current Power handling capability is medium Control circuit is simpler than that for thyristor circuits but more complex than that for MOSFET circuits Majority are NPN transistors but PNP transistors are also used
Power MOSFETs	D G S Drain Gate Source	Power-handling capability is low but parallel drive is easy High frequency switching is possible (e.g. as high as 1 MHz) Simple interfacing with a microprocessor or ICs
Thyristors	Anode Gate Cathode	Power-handling capability is as high as a diode Suitable for high-power applications Spacious commutation circuits are needed, self-turning-off capability is not available integrally itself Switching frequency is lower than that of bipolar transistors

Recently introduced devices are the SIT (static induction transistors), SI thyristor, and IGBT (insulated gate bipolar transistor).

We shall first examine the differences between the diode, bipolar transistor, MOSFET, and various thyristors, since they are regarded as the basic switching devices in terms of device constructions and basic functions.

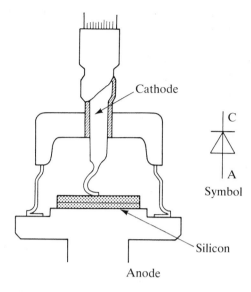

Fig. 2.3. Cross-sectional view of power diode.

2.3 Diodes and PN junctions

First, let us study the basic characteristics of a PN junction. Figure 2.3 illustrates the cross-sectional view of a power diode having one PN junction. A diode is a single crystal of silicon, with one side of the silicon doped with P-type impurity atoms and the other side with N-type impurity.

It is known that a PN junction has the property of a rectifier, i.e. it permits a current to flow in one direction but blocks an opposite current. Rather than physically detailed discussions, a simple but useful explanation is given in Fig. 2.4 of how a PN junction exhibits the property of a rectifier. When a positive potential is applied to the anode with respect to the cathode, the PN junction is forward biased and can carry a current. On the contrary when the PN junction is reverse biased, i.e. when a negative potential is applied to the anode with respect to the cathode, the diode blocks the flow of current. Thus, forward biasing is equivalent to the ON state and reverse biasing to the OFF state. Consequently, when a diode is connected as in the circuit shown in Table 2.2, the diode will carry a current in the positive half-cycle of the applied a.c. potential and will block the current in the negative half-cycle.

Figure 2.5 illustrates the current-versus-voltage characteristics of two typical types of diodes. As illustrated by the solid curve, in a normal diode, the current can flow only when a positive potential higher than 0.6 V is

Table 2.2. Classification of semiconductor devices based on the PN structure.

Devices	PN structure	Symbols	Basic functions
Diode	Anode — P N — Cathode	A ▷|– C or K	e_L
NPN transistor	Collector — N P N — Emitter; Base	C B E	e_i, E, e_o; E
PNP transistor	Collector — P N P — Emitter; Base	C B E	e_i, e_o; $-E$, E
Thyristor	Anode — (four layers) — Cathode; Gate	A ▷|– C or K; G	i_g, e_L; Firing angle α

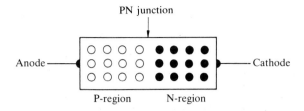

(1) *When no potential is applied to the diode*, holes or positively charged particles are free to move in the P-region, and electrons or negative particles are free to move in the N-region.

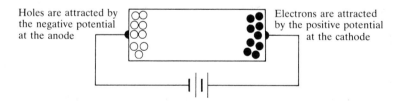

(2) *When forward biased*, i.e. when a potential is applied as above, holes and electrons drift toward the PN junction owing to the electrical field in each region. In the junction, holes and electrons combine with each other to become neutral and disappear. However, new holes are supplied from the anode and new electrons are supplied from the cathode. Thus continuous flows of both sorts of particles are maintained; this is an electric current.

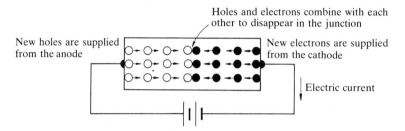

(3) *When reverse biased*, i.e. when a potential is applied as above, holes are attracted by the negative potential at the anode and they are absorbed by it, and electrons are attracted by the positive potential at the cathode and absorbed by it. Thus all the charge carriers are evacuated from the diode; no current will flow.

Fig. 2.4. How a PN junction works as a rectifier.

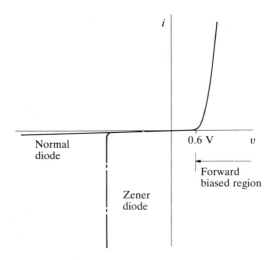

Fig. 2.5. Current-versus-voltage characteristics of diodes. The solid curve is for a normal rectifier diode, and the chain curve is for the Zener diode having constant-voltage characteristics in the reverse-biased region.

applied to the anode with respect to cathode, while in the reverse-biased region only a negligible current can flow. As shown by the chain-line curve, however, in a so-called Zener diode a kind of breakdown can take place due to the tunnel effect at a relatively low reverse potential known as the Zener potential. In this breakdown region the device shows very good constant-voltage characteristics that are utilized in a voltage stabilizer circuit. In some diodes, the Zener potential is as low as 3 V but in others it is as high as 20 V.

2.4 Bipolar transistors

A bipolar junction transistor has two P N junctions in either the P–N–P or the N–P–N construction. No matter which type it belongs to, the central region sandwiched by the two junctions is called the base and denoted by B. One of the two remaining regions is larger than the other, as seen in the 'triple diffusion planar' transistor illustrated in Fig. 2.6; this region is called the collector and denoted by C. The rest is the emitter (E).

2.4.1 *Principles of transistors*

Figure 2.7 illustrates an N P N transistor connected in the 'common emitter' scheme. In this type of connection, the base is used for the input terminal and the collector for the output terminal, while the emitter is common both to the

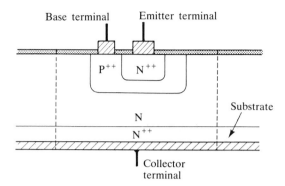

Fig. 2.6. Cross-sectional structure of one unit of a triple-diffusion planar transistor fabricated on a highly doped substrate denoted by N^{++}. It is seen that the collector area is much larger than the emitter. The SiO_2 film covering the boundary of the PN junctions between collector and base is to increase the potential it can withstand. Hundreds or thousands of such units are fabricated on a substrate and they are connected in parallel to be able to handle high currents.

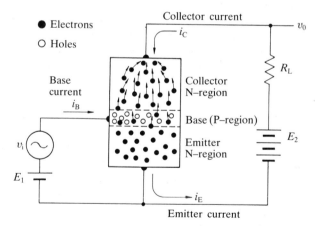

Fig. 2.7. Movement of charge carriers in the common-emitter connection. Most electrons injected from emitter to base region travel by diffusion towards the collector region to produce the collector current. Some electrons recombine with holes in the base. To supply the holes lost in the base, a current flows into the base.

input and output stages. A relatively low forward potential E_1 plus an alternating signal potential is applied between the base and emitter. The d.c. potential E_2 on the other side is higher than E_1. Hence a reverse potential is applied across the PN junction between the collector and base. Since the PN junction between B and E is forward biased, free electrons enter the base region from the emitter.

It is very important that the base region is made so thin that most electrons coming from the emitter penetrate through the base and enter the collector region. In this region the electrons are accelerated towards the collector terminal by the reverse potential of E_2. When signal potential v_i is higher, more electrons will travel from the emitter to the collector region producing more current. On the other hand, when v_i is negative enough to reverse-bias the base–emitter junction, no electrons will travel towards the base or the collector; no collector current will flow.

As stated before, the base region is so thin that most electrons injected from the emitter to the base enter the collector region, having no opportunity

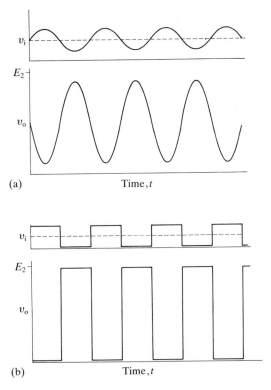

Fig. 2.8. Relationship between the input signal v_i and output signal v_o; (a) the case of sinusoidal input signal; (b) the square-wave input signal.

of combining with a hole. However, the probability of recombination between a hole and an electron in the base is not absolutely zero. Some electrons and holes are lost owing to recombination. To supply the holes to the base region, a current flows from the input power supply (E_1 and v_i) towards the base; this is the base current.

When the collector current varies with time, the potential across the load resistor R_L will vary, too. Figure 2.8 illustrates two cases of relationships between v_i and v_o; one is for a sinusoidally varying input signal and the other is a square-wave input signal. It is seen that a transistor can be used as either a signal amplifier or a as solid-state switch.

When transistors are used as an amplifier to drive a motor, the simple common emitter connection is not employed because in this connection parameters like the current amplification factor differ from transistor to transistor and are strongly affected by temperature. Practical use of transistors in the linear region will be discussed in Chapter 5.

In power electronics, bipolar transistors are used more often as solid-state switches. The bipolar transistors feature high current density per unit area of the semiconductor material. Detailed techniques for this implementation will be discussed in this chapter and other chapters.

2.4.2 *Collector characteristics*

When we discuss the physical properties of a transistor, we usually deal with them in terms of potentials applied to the junctions. For discussing a transistor as circuit element, however, it is convenient to analyse its function in terms of current parameters. Figure 2.9 is a graph that represents the relation

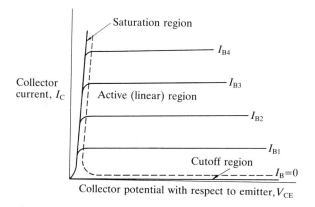

Fig. 2.9. Collector characteristics of a bipolar transistor in the common-emitter connection.

between the collector currents and the collector-to-emitter potential with the base current as a parameter.

The transistor is thought of as having three distinct regions of operation:

(1) active or linear region;
(2) saturation region; and
(3) cutoff region.

These characteristic regions are physically specified in terms of the bias potentials applied to the two junctions as explained in Table 2.3.

Table 2.3.   Relation between biasing states and characteristic regions

	Active	Saturation	Cutoff
Emitter-to-base junction	Forward	Forward	Reverse
Collector-to-base junction	Reverse	Forward	Reverse

When no current is supplied to the base, only a negligible collector current flows; this is known as the cutoff region. When a base current is provided, the transistor is in either the saturation or the active region. In the saturation region, curves for different base currents are almost aligned, starting from the origin. In the active region, the curves for various parameters are nearly parallel to the horizontal axis, branching off from the saturation region.

Figure 2.10 explains the similarlity between a transistor switch and a mechanical switch. When no current is supplied to the base, the transistor acts as an open switch; when sufficient base current is provided, it works as a close switch.

The ratio of collector current I_C to base current I_B in the active region is called the current amplification factor and indicated by h_{FE}:

$$h_{FE} = \frac{I_C}{I_B}.$$

$$(2.1)$$

This parameter is not constant but varies with the applied potential, current, and temperature. As a general tendency, the smaller the transistors the larger

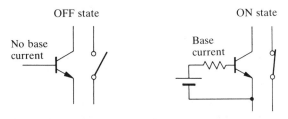

Fig. 2.10. Correspondence between a mechanical switch and a transistor switch.

the h_{FE}. In power transistors handling more than 30 A, typical values of h_{FE} are as small as 10 to 20.

When a transistor is used as an amplifier, as in a linear type d.c. servo-amplifier dealt with in Chapter 5, it must be operated in the active region. However, when used as a switching device, for example in a pulse-width modulation (PWM) servo-amplifier, transistors are to be brought into the saturation region for the ON state and the cutoff regions for the OFF state.

2.5 Power MOSFETs

There are several types of field-effect transistors. The metal–oxide field-effect transistor or MOSFET used as a switching element in power-electronics circuits has a typical construction such as shown in Fig. 2.11(a). The gate is the control terminal, which corresponds to the base of the bipolar transistor. One of the features of the MOSFET is that the gate is isolated by an insulator film from the source and drain, which correspond to the emitter and collector in the bipolar transistor, and hence no appreciable current flows in the gate terminal. The drain electrode is in contact with the N-type substrate, and the source metallization is in contact with both an N-type and a P-type region.

2.5.1 Fundamental properties of MOSFETs

The construction we shall discuss here is the 'N-channel' MOSFET. How channels behave in this construction can be explained using Fig. 2.11 as follows. When no potential is applied to the gate with respect to the source, the N-regions of the source and the drain are separated by the P-region beneath the gate as shown in (a). In this state no electrons can travel from source to drain even if a positive potential is applied between the source and drain. When a positive potential is applied to the gate as in (b), how-ever, the holes in the P-region beneath the gate are depleted owing to the electrical field. This results in increase of the N-region and shrinkage of the P-region to form an N-channel through which current can flow from drain to source.

Thus, while in a bipolar transistor there are two PN junctions in the main current path, there are no PN junctions in the current channel of a MOSFET. Hence, the MOSFET is a monopolar device.

The other type of MOSFET is known as the P-channel MOSFET. To close the P-channel type, a negative potential must be applied to the gate with respect to the source. Figure 2.12 shows the difference in the symbols and how to use these two devices.

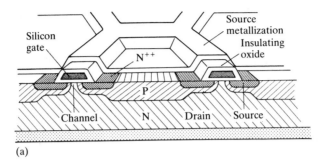

(a)

(a) Cross-section of a MOSFET when no potential is applied. Channels are closed by the P-region.

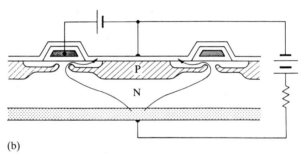

(b)

(b) When a positive potential is applied to the gate with respect to the source, the channels are opened.

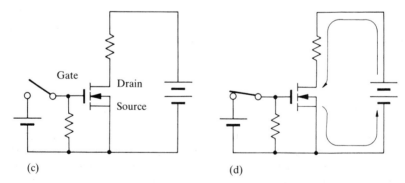

(c)   (d)

(c) When no potential is applied to the gate, the MOSFET does not carry a current.

(d) When a positive potential is applied to the gate with respect the source, a current can flow from drain to source.

Fig. 2.11. Typical construction of MOSFETs and their principle of operation: (a) cross-sectional view of a MOSFET named HEXFET; (b) how N-channels paths are opened to provide current paths when a potential is applied to the gate with respect to the source. Switching actions are illustrated in (c) and (d).

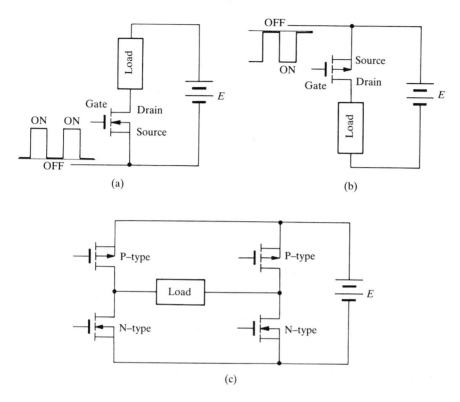

Fig. 2.12. How to use N-channel and P-channel types of MOSFETs. (a) Unipolar drive using an N-channel type; (b) unipolar drive using a P-channel type; (c) bridge configuration using two pairs in complementary connection.

2.5.2 *Characteristics of MOSFETs*

The characteristics of the typical power MOSFET are shown in Fig. 2.13(a) and (b). The relation between the drain current and the gate-to-source voltage is given in (a). This figure reveals that as the gate-to-source voltage is increased from zero, the drain current initially does not increase significantly. Only when a certain threshold voltage has been reached does the drain current start to increase appreciably.

Figure 2.13(b) shows the relation between the drain current and drain-to-source voltage with the gate voltage as the parameter. In the 'constant-resistance region', which corresponds to the saturation region in a bipolar transistor, the voltage increases with the current almost proportionately. In practice, the resistance increases at higher currents; at a certain current level, however, a channel pinchoff effect is reached within the device, and the

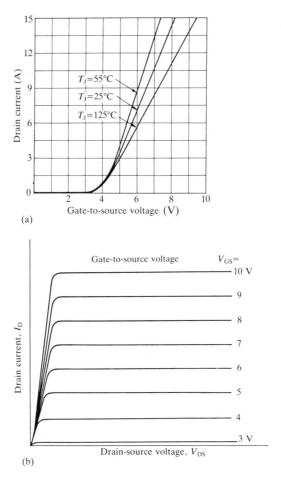

Fig. 2.13. Characteristics of MOSFETs of the enhancement type: (a) drain current versus gate potential; (b) drain current versus drain-to-source voltage with gate potentials as parameter.

operating characteristics move into a constant-current region. As we saw in the preceding sections, the bipolar transistor also exhibits similar collector characteristics. It does not, however, show a truly resistive effect in its saturation region, nor does it exhibit nearly such well-regulated constant-current characteristics in the linear region. It should be emphasized again that the power MOSFET is essentially a voltage-controlled device in contrast to the bipolar transistor which is a current-driven device.

The on-resistance, or the resistance in the constant-resistance region, is obviously an important parameter, because MOSFETs are to be used in this region for switching applications and this parameter determines the drain current under given heatsink conditions.

MOSFETs are superior in switching time to bipolar transistors of the comparable size. The response time of MOSFETs is primarily determined by the capacitance between the gate and source. Although in a stationary condition the input current is only the leakage current and is extremely low, charging and discharging currents flow for a certain period at turning-on and turning-off. This affects the switching speed.

2.6 Thyristor group

There are several types of thyristors, and three types mentioned in Section 2.2 are thought of as the main devices.

2.6.1 *Comparison of reverse-blocking thyristor and a diode*

The standard type of thyristor, which is known as the reverse-blocking thyristor, has three PN junctions and is equipped with three terminals: anode (A), cathode (C or K) and gate (G). This type of thyristor is quite different from a bipolar transistor or a MOSFET in its way of controlling current. Comparison with a diode is more suitable in explaining the basic function of a thyristor. In the figure for the thyristor in Table 2.2, the device starts to carry a current when a trigger signal is applied to its gate terminal.

Thus a thyristor is turned on (or fired) by a trigger current that flows from the gate (G) to cathode (C). Before the trigger signal is applied, the thyristor does not fire even when a voltage is applied to the anode (A). Note that a diode carries a current when a positive voltage is applied to its anode with respect to the cathode. Thus by using a thyristor and a single-phase supply we can control the average d.c. current in the load by the adjustment of timing when applying the gate signal. Since the thyristor is a type of 'controlled rectifier' made of silicon, this device is also known as silicon controlled rectifier or shortly SCR.

Note that the gate of the thyristor has no function of turning off the anode current. In the circuit of Table 2.2, the thyristor turns off automatically when the current falls to zero due to the potential reverse.

2.6.2 *Means for turning off the thyristor*

When a thyristor is used in a d.c. circuit, a means for turning off the device is essential. In the simple circuit shown in Fig. 2.14(a), the device can close the circuit by firing the thyristor by applying a trigger current to its gate. However, there is no means to stop the current unless some auxiliary method

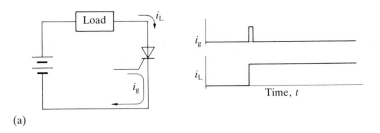

(a)

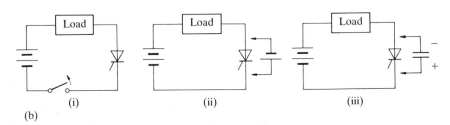

(b)

Fig. 2.14. Basic function of the thyristor. (a) A thyristor used in a plain d.c. circuit can be fired but not turned off by the trigger operation. (b) Methods of turning off the thyristor.

is provided. The problem is how to turn off the thyristor: the gate is of no use for this.

There are two methods of turning-off, as shown in Fig. 2.14(b); one is to use another switch, for example a mechanical switch, only for opening the circuit (see part (i)). The other is to apply a reverse potential across the anode and cathode, for example by using a battery as shown in part (ii). However, these two are not really practical. A practical method is to use a charged capacitor as shown in (iii). If the reverse potential is applied between anode and cathode for a time as short as 10 μs or a little longer by such a capacitor, the thyristor can be turned off. A practical example will be shown in Section 2.11.

2.6.3 GTO

The GTO or gate turn-off thyristor is a type of thyristor whose gate can be used also for turning off the anode current. As illustrated in Fig. 2.15, when a current is drawn from the gate the device turns off.

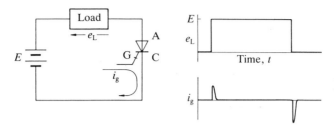

Fig. 2.15. Basic function of a GTO thyristor.

2.6.4 *Triacs*

The triac is a bidirectional thyristor. As this device is used mostly in a.c. circuits, it is also known as the a.c. thyristor. The junction structure, symbol and basic operation are illustrated in Fig. 2.16. A triac can be turned on either by supplying a current to the gate or by drawing a current from it when a negative or positive potential is applied across the terminals 1 and 2. By varying the timing of applying trigger signals to the gate, we may control the a.c. current at the load. If the load is an incandescent (filament) lamp, the brightness can be controlled, and if the load is a single-phase motor to drive an electric fan, the wind flow can be adjusted.

2.7 **New devices**

(1) *Static induction transistor (SIT).*   Another remarkable power device is the static induction transistor (SIT), which was invented by J. Nishizawa and has been developed in Japan. The basic structure is illustrated in Fig. 2.17(a). The gate consists of highly doped P-region grids in the N-region. When no potential is applied to the gate, currents can flow through the channels through the gate grids. When a negative potential is applied to the gate, the P-region expands to shut the channels and blocks the current. The drain current versus drain-to-source voltage characteristics are as shown in Fig. 2.17(b). Thus, the SIT is basically a normally-on device, but a normally-off type similar to most power MOSFETs is available too.[1] The normally-on SIT features high-speed operation and high-power output capability.

(2) *Insulated-gate bipolar transistor (IGBT).*   The bipolar transistor features high current density, while the MOSFET is advantageous in being a high-response voltage-operated device and more compatible with interfacing to a microprocessor. The insulated-gate bipolar transistor or shortly IGBT is

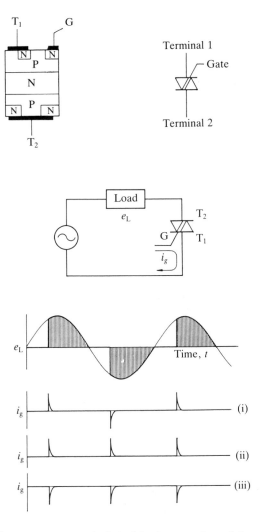

Fig. 2.16. Junction structure, symbol, and basic operation of the triac. There are three possible ways of applying trigger signals.

a monolithic device formed by combining the merits of these two different types of devices. The equivalent circuit of the IGBT is shown in Fig. 2.18.

As seen above, there are many types of power devices and new devices are continuously being developed. Temple[2] discussed the evolution of various power devices in terms of the number of PN junctions in the ON-state current path and the source of the charge that controls this current.

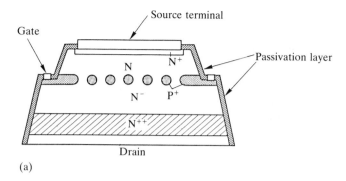

(a)

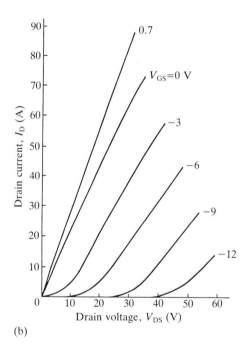

(b)

Fig. 2.17. (a) Basic structure and (b) drain current versus drain-to-source voltage characteristics of the SIT.

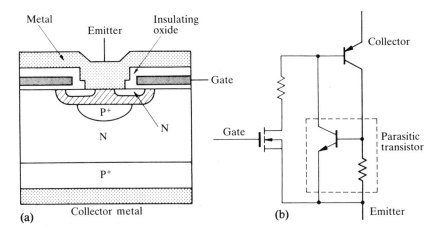

Fig. 2.18. (a) Basic structure, and (b) equivalent circuit of the IGBT.

2.8 Transistor circuits

There is a relatively long history of the application of transistors as switching elements, and consequently there are several important techniques for using bipolar transistors.

2.8.1 Basic circuit aspect

We shall first consider the basic common-emitter circuit of Fig. 2.19. For the sake of simplicity the load is a resistor R_L that is connected between the transistor's collector and the positive terminal of the power supply. A train of square-wave switching signals is applied to the base via a resistor R_B. The low level of the switching signal is zero volt and the high (H) level is E_1. We expect that the transistor enters the saturation region when the H level signal is applied. If the collector-to-emitter voltage V_{CE} is ignored, the voltage equation in the output side is

$$E_2 = I_C R_L. \qquad (2.2)$$

Hence the load current I_C is

$$I_C = \frac{E_2}{R_L}. \qquad (2.3)$$

This means that the collector or load current is determined by the supply potential E_2 and the load resistance R_L, independently of the transistor para-

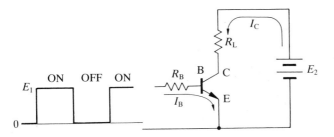

Fig. 2.19. A simple common-emitter connection using an NPN transistor.

meter. But note that the transistor's rated current must be higher than I_C determined by this equation. In order that the transistor is in the saturation region, the base current I_B must be higher than I_C/h_{FE}, which means that

$$I_B > \frac{E_2}{R_L h_{FE}}. \tag{2.4}$$

The h_{FE} parameter varies from transistor to transistor and also with temperature, so the maximum possible base current must be determined using the minimum possible value of h_{FE}. Hence the external resistance R_B must be selected to satisfy the following relation

$$R_B = \left(\frac{E_1 - V_{BE}}{E_2}\right) R_L h_{FE(min)} \tag{2.5}$$

where V_{BE} is the base–emitter forward potential.

When the signal level is at the L level, no base current flows, and consequently the transistor is brought into the cutoff region or the OFF state. For a short time after the signal has fallen to the L level, however, load current flows owing to excess charge stored in the vicinity of the base-to-emitter PN junction.

As shown in Fig. 2.20(a), to make the switching-off time faster in high-frequency applications, the remaining charge carriers must be removed from the base region as soon as the ON period ends. An effective means is always to apply a negative low potential via an appropriate resistor R_1, as illustrated in Fig. 2.20(b). When a positive potential, which must be more effective than the negative bias potential is applied to the base via R_B, it causes the transistor to turn on. When the positive input potential is removed, the excess charges in the base will soon be absorbed to the negative source. A compromise for the sake of simplicity of the circuit is just a resistor R_1 placed between base and emitter as shown in Fig. 2.20(c). The two resistors in this circuit must be selected carefully, because if the value of R_1 is too low, the ON current can

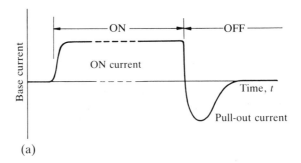

(a)

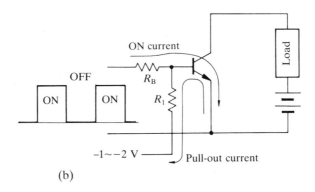

(b)

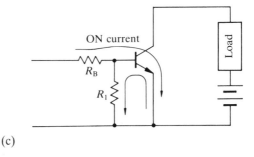

(c)

Fig. 2.20. Quick turning off of a bipolar transistor. (a) Removal of remaining charge carriers from the base region is necessary for quick turning off. (b) An effective method is pulling the base terminal by a negative voltage of 1 or 2 V via a resistor. (c) A simple resistor arrangement by connecting a resistor between the base and emitter to provide a discharge path for the excess carriers.

bypass through it without entering the base. A practical example is seen in Fig. 4.21.

2.8.2 *Darlington connection*

For switching a high current with a small input current, so-called Darlington-connected transistors are used. As shown in Fig. 2.21(a) the emitter current in the first transistor Tr_1, which is much higher than the base current, is the base current to the second transistor Tr_2. Figure 2.21(b) shows a combination of a small PNP transistor and a large NPN transistor, being equivalent to a large PNP transistor. A combination of three transistors can be employed when an extremely high current gain is required.

However, it should be noted that the Darlington scheme increases the saturation voltage in Tr_2. For example, the saturation voltage at a single transistor is 0.4 V or so, but in a double Darlington type it is about 1.3 V.

Figure 2.22 shows a pair of arms of a three-phase bridge modular transistor. As seen in this figure, each monolithic unit consists of a triple Darling-

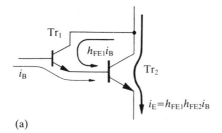

(a)

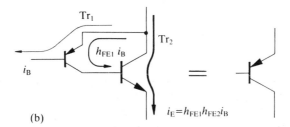

(b)

Fig. 2.21. Darlington connections for increasing effective current amplification factor. (a) Combination of a small NPN transistor and a large NPN transistor, which is equivalent to a large NPN transistor having a large current amplification factor $h_{FE1}h_{FE2}$. (b) Darlington connection as combination of a small PNP and a large NPN transistor, which is equivalent to a single large PNP transistor.

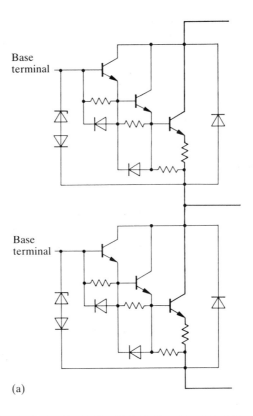

Base terminal

Base terminal

(a)

(b)

Fig. 2.22. (a) Recent power transistor incorporating a current-limiter function, and (b) its module type. (By courtesy of Fuji Electric Co., Ltd.)

ton scheme and a flyback diode, which will be explained in Fig. 2.25. In this modular sheme, a current-limiter function is incorporated using the emitter resistor and the Zener diode placed in the input stage. Two other diodes placed between base and emitter are for quick removal of remaining charge at transient turning-off periods.

2.8.3 Emitter follower schemes

In some applications a motor can be driven as the load of an emitter follower circuit as shown in Fig. 2.23(a). In this case, the switching signal is usually applied to the base without a resistor, but the signals for the H level must be almost the same as the power supply potential E. The emitter follower is used in either switching mode or linear mode as a power amplifier. When the base-to-emitter forward potential, which is normally 0.6 V, is neglected, the load potential is the same as the input signal potential. When signal potential

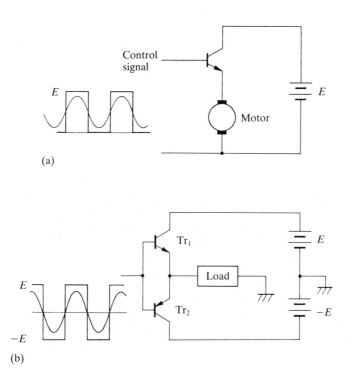

Fig. 2.23. Emitter follower connections where the load is connected to the emitter(s): (a) unipolar emitter follower, and (b) bipolar or complementary type in a dual-supply scheme. The emitter follower has use in either switching or analogue mode.

varies quickly like a pulse, a diode must be placed across the load terminals. The function of this diode is discussed in Fig. 2.25.

Figure 2.23(b) is the complementary emitter follower that is fed from two power supplies E and $-E$. In this scheme, a PNP transistor and an NPN transistor are connected in cascade at the emitters of both transistors, to which the load is also connected. The switching signal, which should be applied to the bases of both transistors, must swing between $+E$ and $-E$. When the input potential is positive, Tr_1 closes and Tr_2 opens. When the input is negative, Tr_1 is open and Tr_2 is closed. Details of the connection between the power stage and a microprocessor will be discussed in Chapter 5.

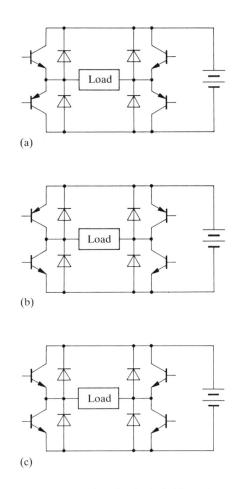

Fig. 2.24. Various cascade connections in the H-bridge scheme: (a) complementary emitter follower used in the bridge scheme; (b) complementary collector follower; and (c) common emitter using NPN transistors only.

2.8.4 *Cascade scheme in the H-bridge*

Figure 2.24 illustrates three practical forms of combination of NPN and PNP transistors in the cascade or complementary scheme in the so-called H-bridge configuration. Using PNP transistors for both source and sink is very rare. The first scheme is the complementary emitter follower discussed above. That using a PNP as the source and an NPN as the sink is suitable for low- and medium-power applications, and an example is shown in Fig. 2.25. The last type employing the same NPN transistor for both source and sink features a wide choice of devices. Interfacing to the control signals will be discussed in Section 2.8.6.

2.8.5 *Flyback and free-wheeling diodes*

In the circuits presented in Fig. 2.24, a diode is placed between the collector and emitter of each transistor. These diodes are needed in many cases to protect solid-state devices switching a current flowing in an inductive load. Let us consider the functions of these diodes in Fig. 2.25(a). Now Tr_1 and Tr_4 are ON, and Tr_2 and Tr_3 are OFF, and a current is flowing in the inductive load as indicated. When Tr_1 and Tr_4 are opened, the load current, which has the tendency to keep flowing in the same direction, will flow through D_2 and D_3 and thus inductive energy is flybacked to the power supply.

There is no problem here whether Tr_2 and Tr_3 are closed at the same time or not. If these diodes are not provided, the OFF transistors will certainly be damaged owing to a high voltage that will build because of a sudden change in the current in the inductive load. When Tr_1 and Tr_4 are turned OFF from ON, diodes D_2 and D_3 will protect these transistors from damage. For this reason, these diodes are called 'flyback diodes' after their function.

Next let us see what will happen when Tr_4 is always closed and Tr_2 and Tr_3 are always open. Tr_1 is used as the switch now. As seen in Fig. 2.25(b_1), a current flows toward right when Tr_1 is ON. When this switch is opened, the load current will continue to flow through diode D_2 as shown in (b_2). It should be noted that the current does not fly back to the power supply but circulates in a 'free-wheeling' manner through the load and circuit elements, which are here D_2 and Tr_4. Because of this function, the diode placed in parallel with the switching element is often called a 'free-wheeling diode'. Whether the diode is for 'flyback' or for 'free-wheeling' depends on how the circuit is used.

In a simple common emitter circuit (c), the diode is always for free-wheeling.

In a MOSFET, a flyback or free-wheeling diode is formed as a parasitic diode as shown in Fig. 2.26. However, it should be noted that this diode cannot be used in certain applications where fast recovery is essential.

To discuss this problem let us again look at Fig. 2.25(a). When transistors' switching states become as (a_1) again from (a_2), diode D_2 and transistor Tr_1

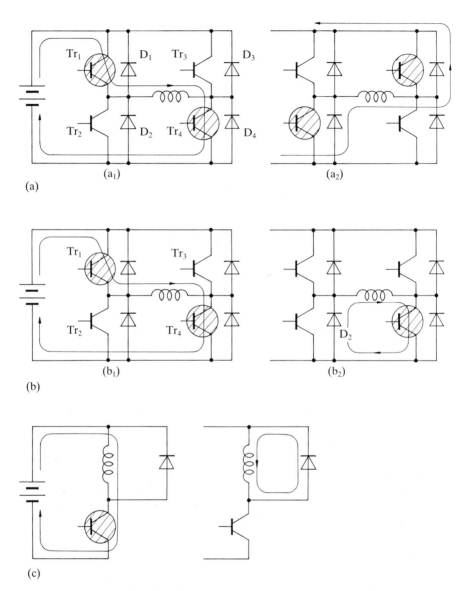

Fig. 2.25. Flyback diodes and free-wheeling diodes. (a) All transistors are switched at the same timing, but Tr_1 and Tr_4 are in the same phase, and Tr_2 and Tr_3 are in the opposite phase. (b) Tr_2 and Tr_3 are always OFF and Tr_4 is ON when Tr_1 is driven in the PWM mode. (c) In the unipolar drive, the diode connected in parallel with the load is always for free-wheeling.

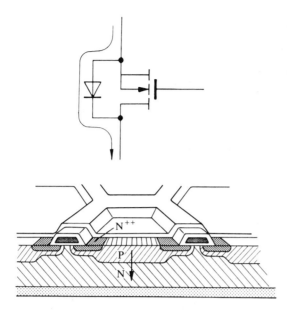

Fig. 2.26. Path for flyback current through parasitic diode in a MOSFET.

can make a short-circuit path, because diode D_2 cannot recover the reverse-blocking property soon enough. Diode D_3 and transistor Tr_4 will provide a short circuit, too. It is very difficult to explain this without referring to solid-state physics based on quantum theory. When the flyback diode needs a long time before recovering the reverse-blocking effect, turning ON at transistors must be controlled so as to take an appropriate time, or a small inductor must be placed between the two transistors in the cascade connection.

If the properties of the parasitic diode in a MOSFET are unsatisfactory, a better diode should be placed across the source and drain terminals. In such an arrangement, the parasitic diode will not work.

2.8.6 *Interfacing to bipolar transistors*

Figure 2.27 is an example of interfacing with control signals for the scheme of PNP source and NPN sink. As is obvious, this circuit features a simple arrangement and is suitable to low- and medium-power applications.

Figure 2.28 illustrates an example of interfacing high-frequency PWM signals to a two-NPN scheme. This uses photocoupler diodes or transistors for electrical isolation between control and power circuits, as illustrated in

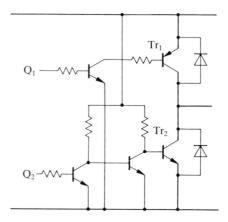

Fig. 2.27. Interfacing for a scheme of source PNP and sink NPN.

part (b). To provide the necessary operating voltage for the interfacing stages, a simple power supply as shown in (c) may be used.

It should be noted that the combination of capacitor C and two diodes seen in Fig. 2.28(b) is for providing a negative potential for quick removal of excess charge from the power transistor bases in order to reduce turning-off time (cf. Section 2.8.1 and Fig. 2.20(b)). When the photodiode (transistor) PT is ON, Tr_4 is closed to provide a base current to the power transistor. At this time Tr_5 is OFF. When the photo device is OFF, Tr_4 is OFF and Tr_5 is ON. At this time Tr_5 works effectively to remove the excess charge from the power transistor, because the emitter potential of Tr_5 is lower than the potential at point A to which the emitter of the power transistor is connected.

2.9 MOSFET circuits

A big difference seen in a MOSFET as compared with a bipolar transistor is that the MOSFET is a voltage-controlled device. This means that when no potential is applied to the gate, the device works as an open switch; when a positive potential is applied to the gate, the device closes and no gate current flows except for a very low transient current to charge the parasitic capacitor between the gate and drain.

MOSFETs are more adaptive to microprocessors than are bipolar transistors, since the output signal can be employed directly as the switching signal of a MOSFET, as shown in Fig. 2.29(a). Figure 2.29(b) is an example of using a comparator in the interface stage. When using the scheme

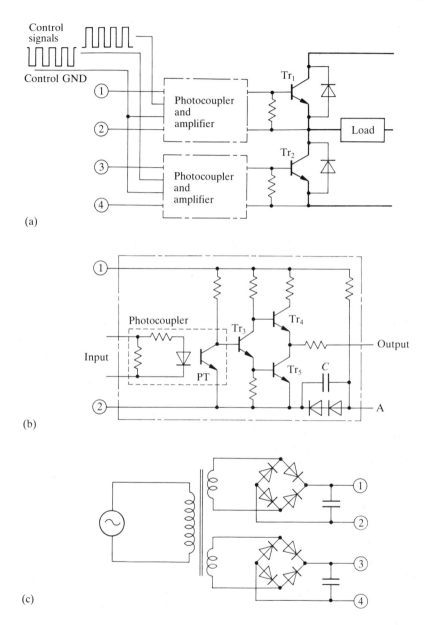

Fig. 2.28. Interfacing for the scheme using NPN transistors for both sink and source: (a) overall configuration; (b) photocoupler and amplifier; (c) isolated power supply for biasing.

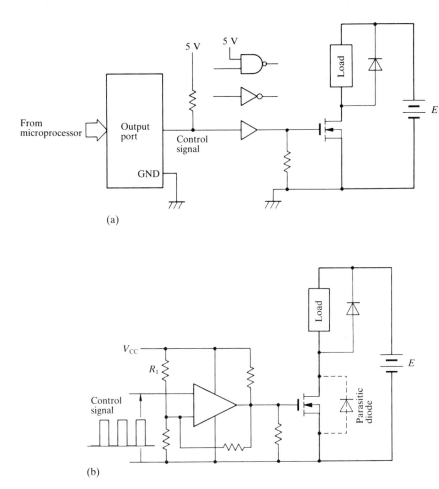

Fig. 2.29. Interfacing to MOSFET unipolar drive: (a) using a logic gate IC; (b) using a comparator.

(a), however, one must confirm on characteristic data like those in Fig. 2.13 that the device can be closed by a gate potential of 5 V or a little lower than it.

Figure 2.30 shows an example of interface between an output port and a pair of push-pull connected arms using two N-channel MOSFETs. Negative bias is not always necessary for quickly turning off the device: compare the case of bipolar transistors shown in Fig. 2.24.

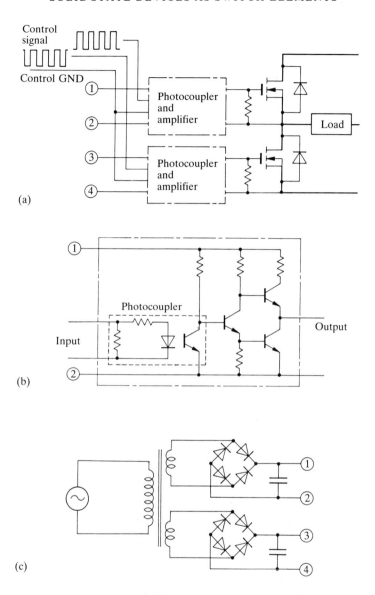

Fig. 2.30. Interfacing using photocouplers: (a) overall configuration; (b) photocoupler and amplifer; (c) isolated power supply.

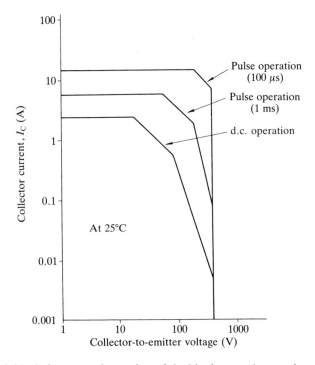

Fig. 2.31. Safety operation region of the bipolar transistor and snubber.

2.10 Safety operation regions and snubber

When using solid-state switches, an auxiliary surge absorber circuit called a 'snubber' is often placed across the device. This is related to the so-called safety operation area of the device. For example, when using bipolar transistors in a power electronic circuit, design engineers should be careful about the current–voltage locus during the turning-off period. When turning off the transistor, the operating point leaves a point at the saturation region for a point in the cutoff region in a short time. However, if the locus of collector current versus collector-to-emitter voltage traces outside the reverse bias safety operating area (or shortly RBSOA), for which data are given in the data sheet of each device (e.g. Fig. 2.31), the transistor will be damaged.

This deviation can occur because of a stray inductance in the circuit: this may raise a potential applied to the collector when the current is suddenly decreased. By placing a snubber, which is a combination of a diode D_2, capacitor C and a resistor R as shown in Fig. 2.32(a), the locus will pass through

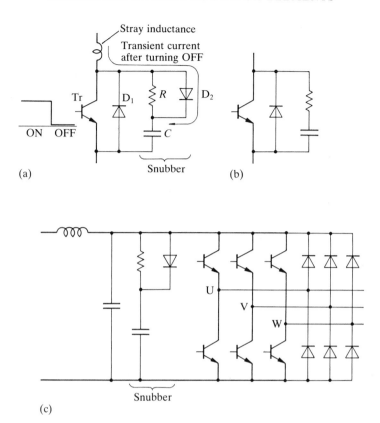

Fig. 2.32. Snubber used in a bridge circuit: (a) function of snubber; (b) simplified scheme; (c) snubber used in a voltage-source inverter.

the lower voltage area. In some simple circuits, diode D_2 is often eliminated, as shown in (b). In a bridge circuit, instead of placing a snubber at each transistor, one snubber placed as shown in (c) can protect all the transistors.

2.11 Direct-current switch using a thyristor

Figure 2.33(a) shows a d.c. switch circuit using a thyristor and incorporating a turning-off mechanism based on the principle explained in Section 2.6.2. At first the thyristor is in the OFF state. When S_1, which may be a small switch, is pushed on for a short time, the device will be turned on. After this, however, another small current I_C will flow through resistor R to charge capacitor C for a short time. After charging, if switch S_2 is pushed, the potential at

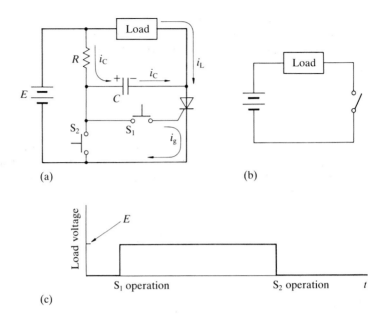

(a)   (b)

(c)

Fig. 2.33. A d.c. switch circuit using a thyristor and a capacitor to drive a turn-off function: (a) circuit configuration; (b) equivalent circuit; switching timing and load voltage.

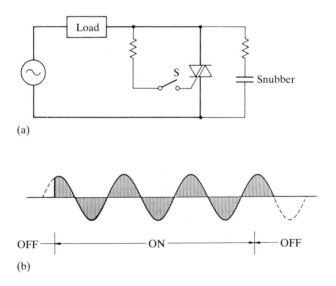

(a)

(b)

Fig. 2.34. (a) A simple triac circuit for switching an a.c. circuit; (b) switching timing and load voltage.

the thyristor's anode suddenly falls to a negative value to turn off the thyristor. These two switches (S₁ and S₂) may be solid-state switches like transistors or thyristors smaller than the main one. Figure 2.33(b) shows the timing chart of the switching operation.

2.12 Alternating-current switch using a triac

Figure 2.34(a) shows a very simple a.c. circuit using a triac. When switch S is closed to provide a trigger current, the triac also turns on, and if S is opened the triac also turns off. The timing relation between switch S and the triac is

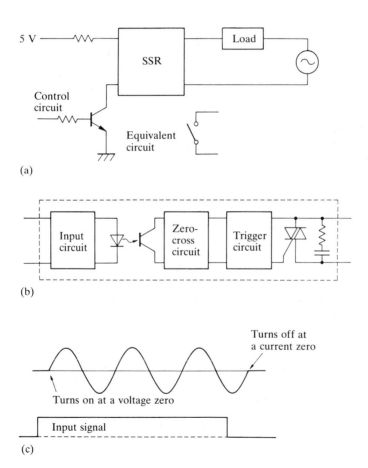

(a)

(b)

(c)

Fig. 2.35. Solid-state relay (SSR). (a) Connection diagram; (b) inside an SSR; (c) zero-cross operation.

illustrated in Fig. 2.34(b). It should be noted that the turning off of the main circuit occurs at the end of the half-cycle in which the switch is opened.

For electromagnetic noise not to be generated, it is desirable that the main circuit is closed at the zero-cross of the applied voltage and that the control signal is electrostatically isolated from the main circuit. Such a device is a kind of module circuit and called the 'solid-state relay' or SSR. A block diagram of an SSR incorporating a photocoupler, a control circuit and a triac, is shown in Fig. 2.35.

Reference

1. Nishizawa, J., Tatsuma, M., and Tamamushi, M. (1987). Recent development of the power static induction transistors in Japan. *Proceedings 14th PCI International Conference, SATECH '87*, pp. 118–132.
2. Temple, V.A.K. (1987). Power device evolution. *Proceedings 14th PCI International Conference, SATECH '87*, pp. 465–475.

3 Converters and phase controllers

The equipment that is used to convert a.c. power to d.c. power is referred to as a converter. The output d.c. voltage is fixed in a simple converter that uses only rectifier diodes. However, if thyristors are used in place of diodes and trigger timing is controlled, the output voltages are adjustable. This technique is referred to as phase controlling and is often employed for speed adjustment of conventional industrial d.c. machines. The phase-control technique can also be used to adjust a.c. voltages for speed adjustment of table fans or brightness of incandescent lamps. In this chapter, converters are dealt with first and follows discussion of a.c. voltage regulation.

3.1 Single-phase rectifier circuits

The simplest converter that uses only one diode is shown in Fig. 3.1. This is known as the half-wave rectifier, because only the positive half-cycles of the single-phase supply voltage are passed to the load.

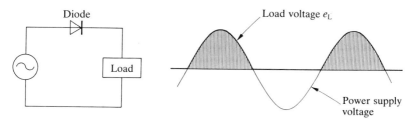

Fig. 3.1. Half-wave rectifier circuit using one diode, and the potential waveform across the load.

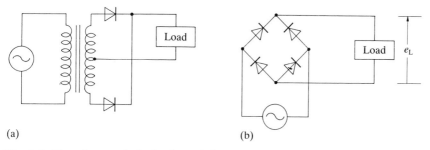

Fig. 3.2. Two forms of single-phase full-wave rectifier circuit: (a) centre-tapped transformer and two diodes; (b) using four diodes.

Basically, there are two forms of single-phase full-wave rectifier circuits, as illustrated in Fig. 3.2. The first type uses a centre-tapped transformer and two diodes, while the second, which is known as the bridge rectifier, is equipped with four diodes and no transformer. The voltage waveform across the load of a full-wave circuit is shown in Fig. 3.3.

Simple LC low-pass filters such as shown in Fig. 3.4 can effectively eliminate the ripple components in the rectified waveform.

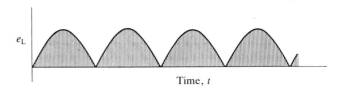

Fig. 3.3. Waveform across the load in a full-wave rectifier circuit.

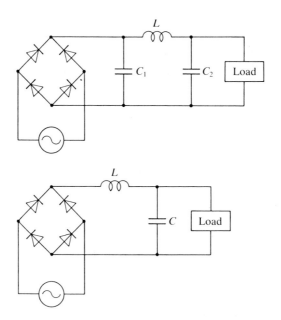

Fig. 3.4. Full-wave rectifier scheme with an LC filter.

3.2 Three-phase rectifier circuits

There are two basic types of three-phase rectifier circuits: the half-wave and
the full-wave schemes. The circuits and waveforms across the load are illustrated in Figs 3.5 and 3.6 for the respective schemes. Obviously, the ripple in
the load potential is less with the full-wave configuration than with the half-
wave one. In the half-wave circuit, in which the primary (or input) side and
the secondary (or output) are electrostatically isolated, a Δ–Y connected
transformer should be used.

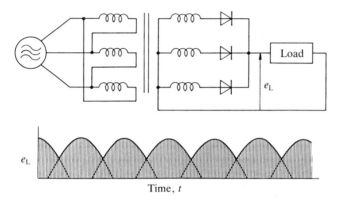

Fig. 3.5. Three-phase half-wave rectifier circuit and the load voltage.

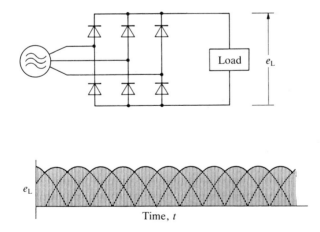

Fig. 3.6. Three-phase full-wave rectifier circuit and the load voltage waveform.

3.3 Voltage regulation using the phase-control technique

By replacing all or some of the diodes in a rectifier circuit by thyristors, we can adjust the output d.c. voltage. For instance, if we use a thyristor in place of the diode in Fig. 3.1, the circuit will assume the fundamental form of the phase-control circuit of Fig. 3.7. The time measured from the start of a positive half-cycle to triggering is called the 'firing angle', denoted by α, and is usually specified in terms of degrees, one cycle of the a.c. voltage being 360 degrees. By varying the firing angle, the d.c. power fed to the load will be continuously controlled. The quantitative relationship between the output potential and the firing angle is shown in Fig. 3.8.

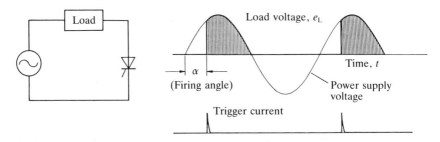

Fig. 3.7. Single-phase half-wave phase-control circuit using a thyristor, and the waveform across the load.

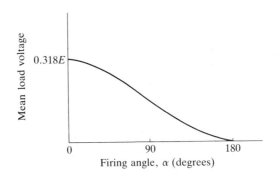

Fig. 3.8. Mean voltage versus firing angle α in the circuit of Fig. 3.7.

3.3.1 Free-wheeling diode

When the load is inductive, there are two possible schemes for phase controllers, as shown in Fig. 3.9. Figure 3.9(a) still has only a thyristor, while the circuit of Fig. 3.9(b) has an additional diode placed in parallel with the load,

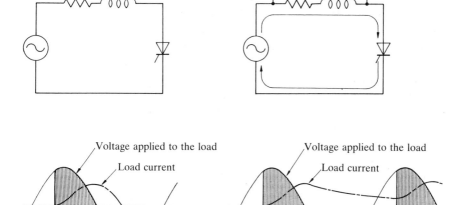

Fig. 3.9. Two schemes of half-wave phase control circuits for an inductor load: (a) using only a thyristor; (b) having a free-wheeling diode to decrease the ripple current in the load.

and this scheme is preferred for obtaining output current with less ripple. The relations between the voltage applied to the load and the current are illustrated in the same figure for the respective cases. In scheme (a) even after the zero-cross of the source voltage, the current will continue to flow, owing to the effect of inductance in the load. As the thyristor is still forward-biased owing to the current, the voltage applied to the load is negative and this voltage causes the current rapidly to fall to zero. If the diode's forward voltage is ignored, the current from start to zero is governed by the following equation:

$$L\left(\frac{di}{dt}\right) + Ri = V_M \sin(\omega t + \alpha). \tag{3.1}$$

The solution of this equation is

$$i = \frac{V_M}{Z}\left[\sin(\omega t + \alpha - \rho) - \sin(\alpha - \rho)\exp\left(-\frac{R}{L}t\right)\right] \tag{3.2}$$

where

$$Z = [R^2 + (\omega L)^2]^{1/2} \qquad (3.3)$$

$$\rho = \tan^{-1}\left(\frac{\omega L}{R}\right). \qquad (3.4)$$

In scheme (b), a difference is seen after the zero-cross, because the load current can be maintained by the path provided by the diode placed in parallel with the load. The thyristor is turned off after the zero-cross as soon as the diode is forward biased. The current before the zero-cross is also expressed by eqn. (3.2). After it, however, the current is governed by

$$L\left(\frac{di}{dt}\right) + Ri + V_F = 0 \qquad (3.4)$$

where V_F is the diode forward voltage.

The solution of this equation is

$$i = B \exp\left(-\frac{R}{L}t\right) - \frac{V_F}{R} \qquad (3.6)$$

where

$$B = \frac{V_M}{Z}\left[\sin\rho - \sin(\alpha - \rho)\exp\left(-\frac{R}{L}\frac{2\pi - \alpha}{\omega}\right)\right] + \frac{V_F}{R} \qquad (3.7)$$

The time before the current falls to zero is much longer than in the case illustrated in part (a). In many cases the next turning-on occurs before the decay of the current. As further cycles are repeated, the average current continues to increase as shown in Fig. 3.10, and eventually reaches a stationary value.

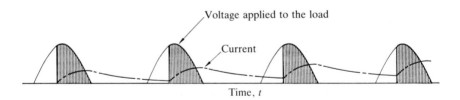

Fig. 3.10. Average current increases with time and eventually reaches a stationary final value for the circuit of Fig. 3.9(b).

Thus, owing only to a diode placed in parallel with the load, the current flowing in an inductive load is made continuous, just as the cogging in a reciprocal engine is reduced by mounting a free-wheel (see Fig. 3.11). For this reason a diode used for such purpose is referred to as a 'free-wheeling diode.'

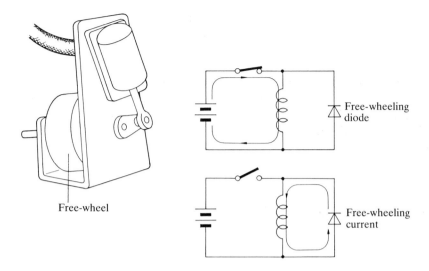

Fig. 3.11. Free-wheel and free-wheeling diode.

3.3.2 *Speed adjustment of a direct-current motor*

An application of the phase-control technique is in the speed adjustment of a d.c. motor. The circuit of Fig. 3.12 is a classic example that employs a Zener diode and a unijunction transistor for adjusting the firing angle. If the d.c. motor is not a permanent-magnet type but a separate-field motor, the field coil is fed from a bridge rectifier.

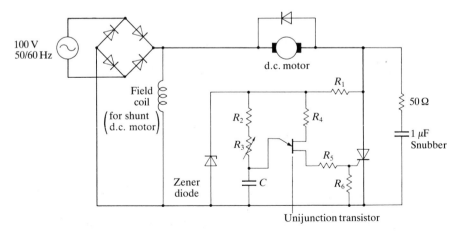

Fig. 3.12. Single-phase full-wave control circuit designed for speed control of d.c. motor of 200 W output power.

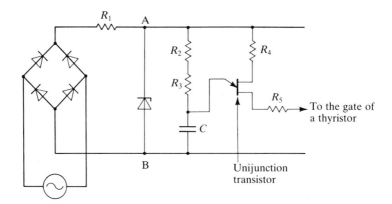

Fig. 3.13. Unijunction transistor oscillator placed in parallel with a Zener diode voltage regulator.

For an explanation of the control circuit, let us use Fig. 3.13 and start with the Zener diode. As explained in Section 2.3, this is a kind of diode having one PN junction used normally in the reverse-biased state. As illustrated in Fig. 3.14, similar characteristics to the ordinary diode are seen in the forward-biased region. However, there is an important difference in the reverse-biased region; as long as the applied potential is lower than the Zener potential, only a negligible current flows. When the reverse potential reaches the Zener level, however, the PN junction will be brought into a breakdown state and a reverse current starts to flow. In this region the potential is constant and independent of the current.

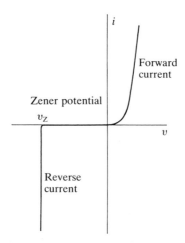

Fig. 3.14. Current-versus-voltage characteristics of a Zener diode.

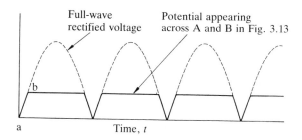

Fig. 3.15. Potential across AB in Fig. 3.13 shows a trapezoidal time variation.

In the circuit of Fig. 3.13 and also in Fig. 3.12, a unijunction transistor circuit is placed in parallel to the Zener diode for generating pulses to trigger the thyristor. In conjunction with the full-wave rectified voltage applied to the circuit, the Zener diode causes the potential across AB in Fig. 3.13 to be trapezoidal as depicted in Fig. 3.15. The potential build-up from point a to point b in this figure is similar to the process of turning on the switch in the circuit of Fig. 3.16. In this circuit the unijunction transistor is represented by an equivalent circuit consisting of a diode and two resistors R_1 and R_2. Resistor R_1 has peculiar current-dependent characteristics, and its resistance decreases as the current increases.

Let us examine the circuit of Fig. 3.16. Before switch S is closed, no potential is applied to the circuit. When the switch is closed, a current will start to flow to charge the capacitor C. Before the capacitor potential or the potential at point A reaches the potential at B, the equivalent diode is reverse-biased and no current will flow through it. However, when the capacitor potential rises enough to bias the diode in the forward polarity, a current will flow through the diode and resistors R_1 and R_4. This current will suddenly reduce

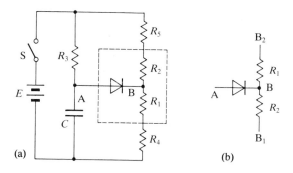

Fig. 3.16. (a) Saw-tooth wave oscillator using a unijunction transistor (UJT). The UJT is represented by an equivalent arrangement in (b).

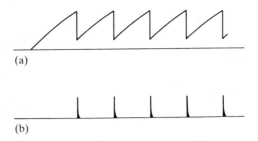

Fig. 3.17. (a) Saw-tooth potential wave observed at point A; (b) current waveform at R_1.

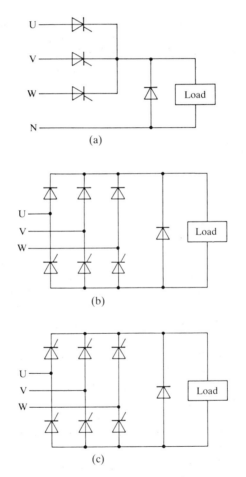

Fig. 3.18. Three basic schemes of three-phase thyristor converters: (a) half-wave converter; (b) full-wave hybrid bridge converter; (c) full-wave six-thyristor converter.

the resistance of R_1. Consequently, the potential at point B decreases and the capacitor is instantly discharged. Since in the circuit of Fig. 3.12 the B_1 terminal is connected to the gate of the thyristor, the discharging current can turn on the thyristor.

When the potential at point B drops to the minimum in the circuit of Fig. 3.16, the initial conditions are recovered and charging of the capacitor will take place again. Thus, as shown in Fig. 3.17, a saw-tooth wave is gene-rated at the capacitor terminal and a train of sharp pulses from terminal B_1 of the unijunction transistor, the frequency being determined by the combina-tion of the capacitance C and the resistance R_3. The firing angle is adjusted by the rheostat R_3 in this circuit.

In this application, however, only the first pulse is needed in a half-cycle of the a.c. potential for firing the thyristor. For this reason, the power source for the control stage is taken from the anode terminal of the thyristor; once the thyristor is turned on, the potential applied to the control circuit is zero, and hence no successive pulses will be generated.

3.4 Three-phase controller

There are basically three types of three-phase thyristor converter circuits as follows:

(1) half-wave converter;
(2) full-wave hybrid bridge converter; and
(3) full-wave six-thyristor bridge converter.

The half-wave scheme can be built by replacing the three diodes of the circuit in Fig. 3.5 by three thyristors; the hybrid bridge type consists of three thy-ristors and three diodes; and the last type uses six thyristors. The funda-mental circuits of these three classes of three-phase converters are illustrated in Fig. 3.18. Figure 3.19 illustrates a comparison of output voltage wave-forms obtained from them. It should be noted that the definition of the firing angle here is different from that for the single-phase controller. The zero of firing angle is defined as the angle that produces the maximum output voltage. In Fig. 3.19(a) it is seen that the firing angle in the half-wave scheme varies up to 150°, and up to 120° in the two full-wave schemes. As for the ripple component included in the output voltage, it is obvious that the half-wave type is the worst and the six-thyristor type is the best. However, the hybrid type contains the same magnitude of ripple as the half-wave type in the region of 60° to 120° firing angle.

A practical circuit of the hybrid bridge type is given in Fig. 3.20, which uses integrated circuits developed for phase-controller use. The internal block diagram of this IC is illustrated in Fig. 3.21.

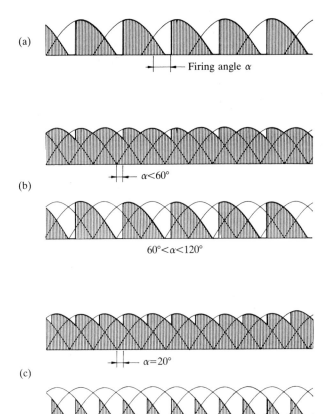

Fig. 3.19. Output waveforms in three different types of three-phase converters: (a) half-wave converter; (b) full-wave hybrid bridge converter; (c) full-wave six-thyristor converter.

3.5 Two-quadrant and four-quadrant operation

When a reactor is connected in series with the load without a parallel diode in a thyristor converter, so-called two-quadrant operation is available. This can be understood using Fig. 3.22. Single-phase bridge and three-phase bridge schemes are shown in (a) and (b), respectively. Here the load is a combination of a reactor and a d.c. power supply. In practice the d.c. supply can be a d.c. motor that can also be operated as a d.c. generator.

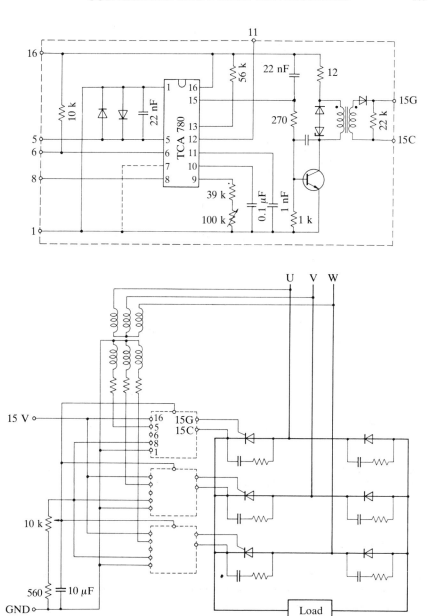

Fig. 3.20. Full-wave hybrid converter using phase-controller chip TCA 780.

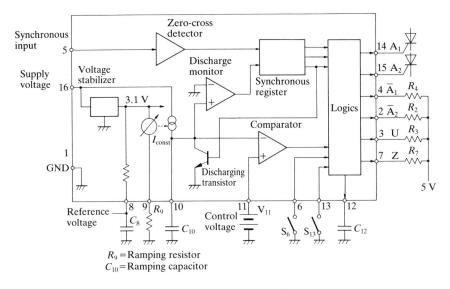

Fig. 3.21. Phase controller chip TCA 780: block diagram and pin arrangement.

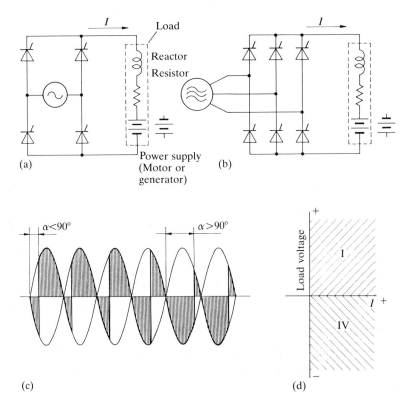

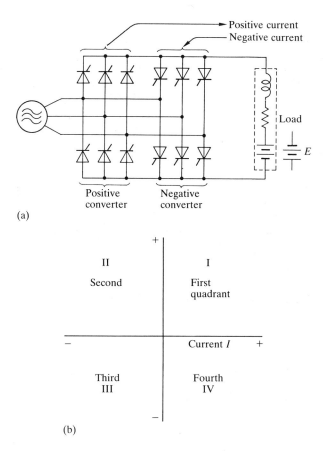

(a)

(b)

Fig. 3.23. Four-quadrant operation using two sets of converters: (a) dual converter scheme; (b) four-quadrant states.

Figure 3.22(c) illustrates the relation between the firing angle and the voltage waveform applied to the load in the single-phase bridge converter. When the firing angle is less than 90°, the mean voltage is positive. This is the first-quadrant operation shown in Fig. (d). However, when the firing angle is greater than 90°, the mean potential is reversed, while the current direction remains the same. The operational state is now in the fourth quadrant. In the first quadrant the electrical power flows from the a.c. supply to the load, while in the fourth quadrant the power is fed back to the a.c. supply from the

Fig. 3.22. Two-quadrant operation of a converter using thyristors: (a) single-phase bridge; (b) three-phase bridge; (c) relation between the firing angle α and voltage waveform; (d) two-quadrant states. The battery in the load may be a power supply, a motor or a generator.

load. Hence, when the load is a rotating d.c. machine, it must be working as a d.c. generator.

In order for four-quadrant operation to be available, two sets of converters must be used as shown in Fig. 3.23. The positive converter supplies a positive current, governing the first and fourth quadrants. A negative current is provided by the negative converter that covers the second and third quadrants. In Chapter 7 the concept of four-quadrant operation of a d.c. machine is expanded to the drive of a three-phase induction motor, and it will be discussed again in Chapter 9.

3.6 Alternating-current voltage regulator using phase-control technique

In the phase control circuits discussed in the preceding sections, the output was a pulsating or smoothed d.c. voltage. The phase-control circuit without rectification is a sort of a.c. voltage regulator.

3.6.1 *Two fundamental types*

Figure 3.24 illustrates two schemes of voltage regulator; (a) uses two thyristors, while (b) employs only one triac as the switching device. Figure 3.25 illustrates how the output potential reduces as firing angle is retarded, and

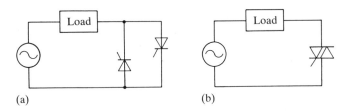

(a) (b)

Fig. 3.24. Two basic schemes of voltage regulator: (a) using two thyristors; (b) using one triac.

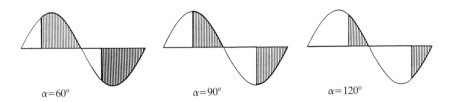

$\alpha=60°$ $\alpha=90°$ $\alpha=120°$

Fig. 3.25. How the load voltage reduces with firing angle in circuits of Fig. 3.24.

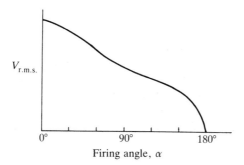

Fig. 3.26. Plots of the root-mean-square voltage as a function of firing angle.

Fig. 3.26 shows the relation between the root-mean-square voltage and firing angle.

Alternating-current voltage regulators of these types are suitable for economical speed control of single-phase induction motors—for example, for the motor used in a table fan. Universal motors, which are also known as a.c. series motors, are also suitable loads of the phase controllers. If the load is an incandescent lamp and its brightness is to be continuously varied, the circuit may be referred to as a dimmer.

3.6.2 Dimmer using two thyristors

Figure 3.27 is an example of dimmer using two thyristors as the main switching elements and a unijunction transistor in the control stage. It is seen that the circuit is not very simple because a transformer having two secondary coils is needed for transmitting pulses to the two thyristors.

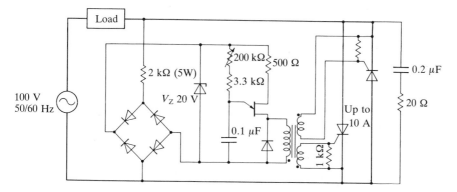

Fig. 3.27. A voltage-regulator circuit using two thyristors with the following rating: power supply; 100 V, 50/60 Hz, and maximum current; 3 A (r.m.s.).

3.6.3 *Dimmer using one triac*

The circuit of Fig. 3.28, which uses one triac and one trigger diode instead of two normal thyristors and a UJT, is much simpler and more economical. The trigger diode has a simple NPN structure and has current-versus-voltage characteristics as shown in Fig. 3.29. When the voltage applied across the two terminals is lower than the break-over voltage V_{B1} or V_{B2}, the current is very small. However, when the applied voltage reaches this critical value the trigger diode is brought into the 'breakdown' state, in which the voltage drops and a large current can flow. This characteristic behaviour is utilized for triggering the triac.

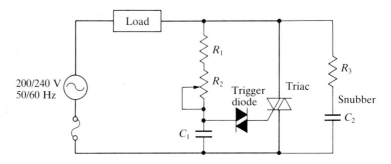

Fig. 3.28. A simple voltage-regulator circuit using a triac and a trigger diode.

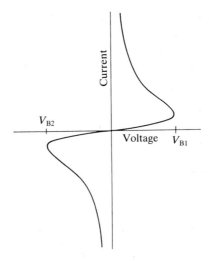

Fig. 3.29. Current-versus-voltage characteristics of a trigger diode.

Let us consider the case in which a positive half-cycle starts and capacitor C_1 is going to be charged via R_1 and R_2. The higher the resistance of R_2, the longer the time needed for the capacitor voltage to reach the breakdown potential of the trigger diode. At the critical potential, the trigger diode breaks down and capacitor C_1 discharges to provide the triac trigger current. A similar procedure will take place in a negative half-cycle. The only difference is the current direction in the triac gate.

Considerations in the choice of solid-state power devices and determination of passive elements (resistors and capacitors) are briefly reviewed here. Selection of the trigger diode is simple, because the options are few. This example employs Toshiba's 1S2093, whose break-over voltage ranges from 26 to 36 V. If one wants to build a dimmer to be used on a single-phase 200 V line, the peak voltage is 283 V. The triac must be able to withstand this potential before firing. If a safety factor of 1.5 is selected, the device's maximum sustainable voltage must be higher than $283 \times 1.5 = 425$ V.

If one is going to control the brightness of five parallel lamps of 100 W, the r.m.s. (root-mean-square) current is 2.5 A. When the same safety factor of 1.5 is to be employed here, the current rating of the triac must be larger than $2.5 \times 1.5 = 3.75$ A (r.m.s.).

The next problem is the time constant or the product of C_1 and R_2. Before

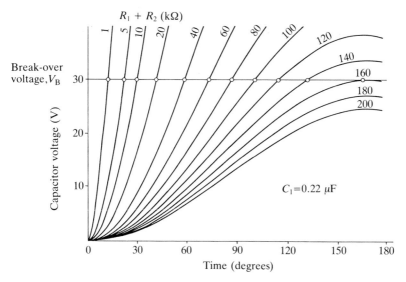

Fig. 3.30. Capacitor voltage-versus-time relationship with the rheostat resistance as the running parameter. When the capacitor voltage reaches the break-over voltage, the triac in the Fig. 3.29 is triggered to fire. In this graph the break-over voltage is assumed to be 30 V.

the triac is fired in a positive half-cycle, the capacitor voltage varies with time according to the following equation:

$$V_C = \sqrt{2}\, V_{\text{r.m.s.}} \sin \phi \left[\cos \phi \exp\left(-\frac{t}{C_1 R_1} \right) - \cos(2\pi f t + \phi) \right] \quad (3.8)$$

where $V_{\text{r.m.s.}}$ = root-mean-square voltage of the power supply
ϕ = phase angle of the impedance = $\tan^{-1}[1/(2\pi f C_1 R_2)]$.

The graph of Fig. 3.30 shows the plots of this equation for a 200 $V_{\text{r.m.s.}}$/50 Hz operation for various values of resistance R_2 with a fixed capacitance of $C_1 = 0.22\ \mu F$. The horizontal time scale is in degrees. When the curves intersect the horizontal line of the break-over voltage, which is assumed to be 30 V here, the trigger diode breaks down and the triac is fired. It is seen that a rheostat of a maximum resistance of 180 kΩ can cover a wide range of values of the firing angle. However, a rheostat of 250 kΩ is recommended for the following reasons:

(1) Owing to variations in capacitors, the effective value can be as low as 70% of the nominal value.
(2) The break-over voltage of the trigger diode may be lower than 30 V.

Resistor R_1 of 1 kΩ is placed in series with the rheostat to make the lowest resistance at this value, which produces a firing angle of about 12°.

The purpose of the snubber in phase-controller circuits is slightly different from that discussed in Section 2.10 for the application of the bipolar transistor. The most evident case is seen in the circuit in Fig. 3.12. In this example the trigger signal is generated when the UJT breaks down. If there is no snubber and the load is inductive, there is a possibility that not enough anode current flows to turn the thyristor on when only a short-period trigger current is applied to the gate. To avoid failure to turn on, the capacitor in the snubber is discharged to supply an enough anode current to the thyristor. The larger the snubber capacitance, the higher the current, which guarantees firing but increases the loss in the snubber resistor.

The situation is similar in the circuits shown in Figs 3.27 and 28. However, there is no simple and practical theory for determining the snubber parameters. It is normal to choose around 50 Ω for R_3 and about 0.2 μF for C_2 for this class of load.

4 Direct-current converters

Conversion of d.c. power from one voltage to another can be implemented with a combination of a reactor or/and capacitor and a solid-state device operated in a high-frequency switching mode. This sort of equipment is called d.c.-to-d.c. converter or just a d.c. converter. The switching technique used in d.c. converters is referred to as PWM, standing for pulse-width modulation. There are two fundamental types of d.c. converter: one is the step-down converter, which produces an output voltage lower than the input one; the other, the step-up type, converts the input voltage to a higher output voltage. Although this chapter will mainly be focused on the theory of d.c. converters and d.c. motor characteristics driven by a step-down converter, some simple but useful circuits will be presented as well.

4.1 Pulse-width modulation and principles of direct-current converters

We shall start our discussion with the basic principle of the pulse-width modulation technique as applied to the step-down converter, and proceed to the step-up converter. This section is based on the theory presented by R.E. Morgan in Reference [1].

4.1.1 *Fundamental step-down converter*

With a step-down converter, the voltage of the input d.c. power is converted to another voltage that is lower than the input, while the output current is higher than the input current. The function of this type of converter is illustrated by the block diagram of Fig. 4.1(a). Ideally, the power loss in the converter is negligible, so that between the input and output we have the following relation:

$$V_o I_o = V_i I_i, \tag{4.1}$$

where V_o = output voltage
 V_i = input voltage
 I_o = output current
 I_i = input current.

It is supposed that, as shown in Fig. 4.1(b), the output voltage is adjustable by the control signal in a certain range from zero to the input level.

In this chapter, in developing the theory, small letters are used for time-

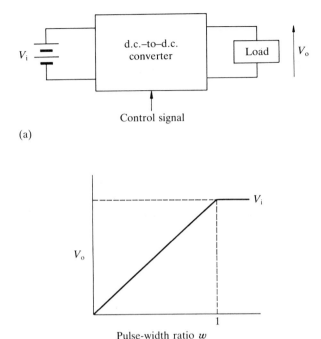

(a)

(b)

Pulse-width ratio w

Fig. 4.1. (a) Block diagram and basic function of the step-down d.c. converter; (b) output voltage as a function of pulse-width ratio.

varying quantities, and capital letters for stationary or time-averaged values.

The elementary principle of the step-down converter is illustrated in Fig. 4.2. Switch S is actually a transistor or other type of soild-state device that can be opened and closed cyclically 1000 times or more in a second. In this circuit, the output voltage is not stationary but varies as a square wave. For the output voltage in such a case one should employ the average value over a repetitive cycle, which is related to the input voltage and the time that the switch is opened or closed as follows:

$$V_o = \frac{V_i T_{on}}{T_{on} + T_{off}}. \tag{4.2}$$

If we use the concept of pulse-width ratio w, defined by

$$w = \frac{T_{on}}{T_{on} + T_{off}}, \tag{4.3}$$

eqn (4.2) is written as

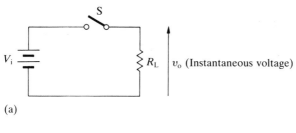

(a)

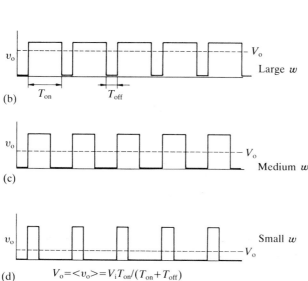

(b) T_{on} T_{off}

(c)

(d) $V_o = \langle v_o \rangle = V_i T_{on}/(T_{on}+T_{off})$

Fig. 4.2. Elementary d.c. converter (a) and its principle, known as the fixed-frequency type, in which the switching frequency is taken as constant.

$$V_o = wV_i. \tag{4.4}$$

As w varies from 0 to 1, the output voltage is ideally a linear function of w, as illustrated in Fig. 4.1(b).

Figures 4.2(b)–(d) show how the load or output waveforms vary as the output voltage is increased, with the pulse-width ratio in a fixed switching mode. In some circuits, the frequency may be varied while the ON time is kept constant as shown in Fig. 4.3, in which the output voltage reduces with frequency. For either fixed- or variable-frequency control, the output voltage is zero when switch S is open, while it is equal to the input (supply) voltage when the switch is closed for a time longer than the normal switching cycle.

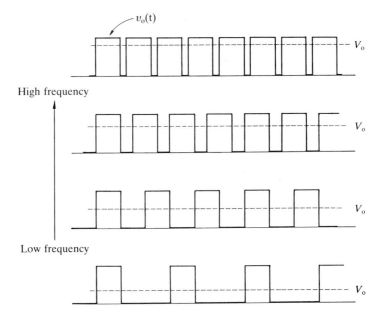

High frequency

Low frequency

Fig. 4.3. Output waveform when the PWM signal is given in the variable-frequency mode.

4.1.2 *Addition of an inductor and a diode to eliminate pulsation of current*

The elementary circuit of Fig. 4.2(a) is only suitable for supplying resistive loads where smooth output current is not required. The scheme shown in Fig. 4.4(a) includes a reactor L and a diode D which are important additions to the elementary circuit to provide a smooth d.c. current to practical loads like an electronic circuit or a d.c. motor, and also to satisfy eqn (4.1). (In Fig. 4.4(a) the load is still just a resistor, for simplicity of explanation.) The thin solid line in the figure indicates the current path when the switch is closed. As discussed in Section 3.3.1, diode D is a free-wheeling diode that provides the path for the load current as indicated by the dotted curve when the switch is open. This diode action permits the use of a simple filter inductance L to provide a smooth d.c. load current for many applications. When the switching frequency is in the kilohertz range, a relatively small inductance is often sufficient to reduce the ripple to a tolerable degree.

Figure 4.4(b) shows the waveform of the potential appearing across diode D. This potential equals the input voltage when the switch is closed, and is zero when the switch is open. As the time average of the potential across

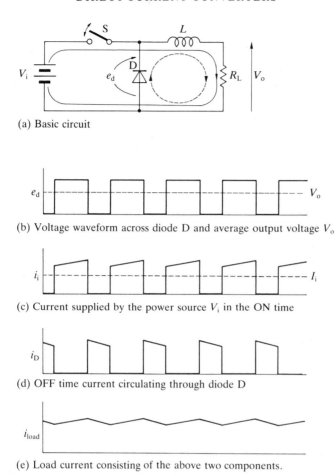

(a) Basic circuit

(b) Voltage waveform across diode D and average output voltage V_o

(c) Current supplied by the power source V_i in the ON time

(d) OFF time current circulating through diode D

(e) Load current consisting of the above two components.

Fig. 4.4. Addition of an reactor and diode to the circuit of Fig. 4.2 to smooth the output current.

inductor L is negligible when it has no resistive component, the output voltage must be the time average of the diode potential. Thus it is seen that eqn (4.2) holds here. Figure 4.4(c) shows the waveform or the output current, and (d) shows the diode current. It is seen in (e) that ripple included in the output current reduces with the switching frequency. However, with this circuit the input current still pulsates.

As illustrated in Fig. 4.5, a capacitor is often used in parallel with the input power source in order to smooth the input current. The time average of the input current is

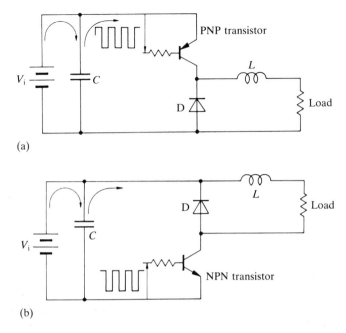

Fig. 4.5. Use of a transistor as the switching device. A capacitor is usually connected in parallel with the input power supply. (a) when a PNP transistor is used in this scheme, the GND terminal is common to the input and output sides. (b) when an NPN transistor is used, the positive terminal of the input battery is common to both sides.

$$I_i = I_0 \frac{T_{on}}{T_{on} + T_{off}}, \qquad (4.5)$$

or, using the pulse-width ratio w defined by eqn (4.3).

$$I_i = wI_0. \qquad (4.6)$$

Thus the input current is lower than the output current, and it is seen that eqn (4.1) is satisfied here.

In Fig. 4.5, the switch is replaced by a transistor; a PNP transistor is used in the circuit (a), in which the negative terminal of the input power supply is common to both the input and output sides, while in (b) an NPN transistor is used.

4.1.3 *Step-up converters*

It is also possible to step up the input voltage with the help of solid-state switch operated in the PWM fashion. The fundamental circuit of the step-up

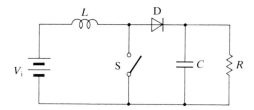

Fig. 4.6. Basic scheme of the step-up d.c. converter.

converter is shown in Fig. 4.6. Reactor L is used to provide a smooth input current. Some ripple component is actually included in the input current, but it is regarded as negligibly small when switching action is repeated at a high frequency.

Now let us see how the load potential becomes higher than the input voltage on this circuit. As shown in Fig. 4.7(a), when the switch is closed, current I_i will flow from the positive terminal of the input battery to the negative one through the reactor and the switch. As shown in (b), when the switch is open, however, this current will go through the diode D and flow into capacitor C and the load. If we denote the diode current by i_D, it behaves as follows:

$$i_D = 0 \quad \text{when the switch is closed (ON)};$$
$$i_D = I_i \quad \text{when the switch is open (OFF)}.$$

Thus, the diode current will pulsate as illustrated in Fig. 4.8.

Capacitor C in Fig. 4.6 is necessary to reduce the ripple in the output voltage and smooth the current supplied to the load. If the capacitor is large enough, the output current will have a negligible ripple component, and will equal the time average of the diode current. Therefore, we obtain the relation

$$I_o = I_i \frac{T_{off}}{T_{on} + T_{off}}. \tag{4.7}$$

We now take conservation of energy into account to obtain the relation between the input and output voltages. Since the input power supply always supplies a constant current I_i, the electrical energy supplied over a repetitive cycle is

$$P_i = V_i I_i (T_{on} + T_{off}). \tag{4.8}$$

On the other hand, the electrical energy supplied to the load over a cycle is

$$P_o = V_o I_o (T_{on} + T_{off}). \tag{4.9}$$

Using eqn (4.7), however, this becomes

$$P_o = V_o I_i T_{off}. \tag{4.10}$$

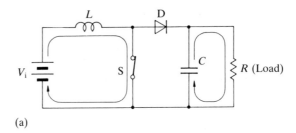

(a)

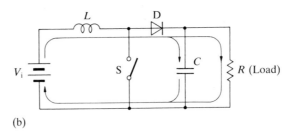

(b)

Fig. 4.7. Current paths in a step-up converter. (a) When the switch S is closed the input current flows through the reactor and switch, and load current is supplied from the capacitor. No current flows through the diode. (b) When the switch S is opened, the input power supply provides current both to the capacitor and load.

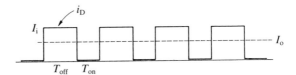

Fig. 4.8. Relation between I_i, i_D, and I_o.

If none of the reactor L, diode D, and capacitor C dissipates energy as heat loss, P_i and P_o must be equal to each other. By equating eqns (4.8) and (4.10) we obtain the relation

$$V_o = V_i \frac{T_{on} + T_{off}}{T_{off}}. \tag{4.11}$$

It is thus seen that the output or load voltage is always higher than the input voltage as long as the switch S is operated at an appropriately high frequency. According to this theory, the output voltage becomes infinite when T_{off} becomes very short. In practice, however, the maximum voltage cannot be infinite, owing to the heat loss in the elements used and the leak current in the capacitor.

Now we raise a problem: Is it possible to design a d.c. converter that can be either step-down or step-up and produces a variable output voltage around the input voltage? This problem will be discussed in Chapter 9.

4.2 Theory of ripple voltage

In designing a d.c. converter, the calculation of the ripple component in the output voltage or current is essential to determine the switching frequency, the condenser's capacitance, and the reactor's inductance in connection with the load. Fundamental theories will be developed here for this problem.

4.2.1 Step-down converter having an RL load

In Section 4.1.2, when developing a theory of adding a reactor and a diode in the step-down converter, it was assumed that there is only negligible ripple component in the output current. In practice, however, some ripple component must be involved in the current as long as a finite inductance is used. We shall, therefore, develop a simple theory for determining the approximate ripple percentage.

We assume that, in Fig. 4.4, the load is purely resistive and consider the load current when the switch is closed. Since the diode is reverse-biased now, the equivalent circuit is as shown in Fig. 4.9(a) and its voltage equation is expressed as

$$L \frac{\mathrm{d}i}{\mathrm{d}t} + Ri = V_i \tag{4.12}$$

The general solution of this equation is

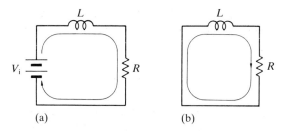

(a)   (b)

Fig. 4.9. Equivalent circuit for the step-down d.c. converter for analysis: (a) for the T_{on} period; (b) for the T_{off} period.

$$i = \frac{V_i}{R} + K \exp\left(-\frac{R}{L} t\right) \qquad (4.13)$$

where K is a constant that is to be determined from the initial conditions. If the current just after the switch has been closed is denoted by I_{min}, K becomes

$$K = I_{min} - \frac{V_i}{R}. \qquad (4.14)$$

As V_i/R is the current when the switch is closed, this is larger than I_{min}, and hence K is negative. Equation (4.13) is therefore written as

$$i = \frac{V_i}{R} - \left(\frac{V_i}{R} - I_{min}\right) \exp\left(-\frac{R}{L} t\right). \qquad (4.15)$$

This is a function that increases monotonically with time as shown in Fig. 4.10.

At the moment the switch is opened the current becomes the maximum, that is

$$I_{max} = \frac{V_i}{R} - \left(\frac{V_i}{R} - I_{min}\right) \exp\left(-\frac{R}{L} T_{on}\right). \qquad (4.16)$$

Next, when the switch is open again in Fig. 4.4, the output circuit is isolated from the input side and the load current will circulate through the diode owing to the free-wheeling effect, as illustrated in Fig. 4.9(b). If we ignore the forward potential in the diode, the voltage equation is given by

$$L \frac{di}{dt} + Ri = 0. \qquad (4.17)$$

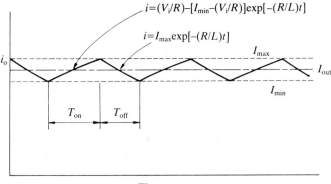

Fig. 4.10. Ripple involved in the output current in the step-down converter.

The general solution of this equation is

$$i = Q \exp\left(-\frac{R}{L}t\right).$$

(4.18)

where Q is a constant that is to be determined from the initial conditions. This equation indicates that the current decreases with time. Since there should not be any sudden change in the current, because of inductance L, the initial value of i, which is Q, must be I_{max}. Hence, we have

$$i = I_{max} \exp\left(-\frac{R}{L}t\right).$$

(4.19)

When the switch is closed again during T_{off}, the current will fall to

$$i = I_{max} \exp\left(-\frac{R}{L}T_{off}\right).$$

(4.20)

In the stationary state this must be equal to I_{min}, as shown in Fig. 4.10, and hence we obtain the following equation:

$$I_{min} = I_{max} \exp\left(-\frac{R}{L}T_{off}\right).$$

(4.21)

By assuming that the switching frequency is so high that the difference between I_{max} and I_{min} is much smaller than the average value, we can derive for the average current the following expression:

$$I_o = \frac{V_i}{R}\frac{T_{on}}{T_{on}+T_{off}}.$$

(4.22)

We shall, however, pay attention to the ripple component in the load current and discuss this problem through an approximate but useful mathematical procedure. When T_{on} and T_{off} are short enough and

$$(R/L)T_{on},\ (R/L)T_{off} \ll 1$$

(4.23)

is satisfied, eqns (4.16) and (4.21) are approximated, respectively, as by

$$I_{max} = I_{min}\left(1 - \frac{R}{L}T_{on}\right) + \frac{V_i}{L}T_{on},$$

(4.24)

$$I_{min} = I_{max}\left(1 - \frac{R}{L}T_{off}\right).$$

(4.25)

The peak-to-peak ripple current can be derived from either of these equations. For example, from the former equation, the difference of I_{max} and I_{min} is

$$\Delta I = I_{max} - I_{min} = \left(\frac{V_i}{L} - \frac{R}{L}I_{min}\right)T_{on}.$$

(4.26)

When the ripple component is much smaller than the average current, I_{min} is approximately equal to I_o given by eqn (4.22), and therefore we obtain

$$\Delta I = \frac{V_i}{L} - \left(\frac{R}{L}\frac{V_i}{R}\frac{T_{on}}{T_{on} + T_{off}}\right) T_{on}$$
$$= \frac{V_i}{L}\frac{T_{on}T_{off}}{T_{on} + T_{off}} \qquad (4.27)$$

Next we shall examine the relation of eqn (4.25). The peak-to-peak ripple is given by

$$\Delta I = I_{max} - I_{min} = I_{max}\frac{R}{L}T_{off}. \qquad (4.28)$$

If we replace I_{max} on the right-hand side with the average load current I_o given by eqn (4.22) as before, we obtain

$$\Delta I = I_{max} - I_{min} = \left(\frac{V_i}{R}\frac{T_{on}}{T_{on} + T_{off}}\right)\frac{R}{L}T_{off}$$
$$= \left(\frac{V_i}{L}\right)\frac{T_{on}T_{off}}{T_{on} + T_{off}}. \qquad (4.29)$$

The peak-to-peak percentage of the current ripple is therefore

$$\text{Peak-to-peak percentage} = 100\frac{\Delta I}{I_o}$$
$$= 100\frac{R}{L}T_{off}. \qquad (4.30)$$

Exercise. Determine the minimum inductance that makes the ripple percentage less than 10%, provided that the switching frequency is 10 kHz and the load resistance is 10 Ω.

Answer and explanation. As the longest T_{off} is 0.0001 s, we obtain

$$100 \times (10/L) \times 0.0001 < 10, \qquad (4.31)$$

from which

$$L > 0.01 = 10 \text{ mH}. \qquad (4.32)$$

Thus, an inductance larger than 10 mH is required.

4.2.2 Step-up converter having a resistive load

Similarly to the case of the step-down converter, the load current waveform of the step-up converter is as shown in Fig. 4.11. Let us consider the current

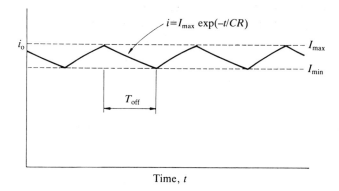

Fig. 4.11. Ripple in the output current of the step-up converter.

after the switch is closed. Since the diode is now reverse-biased, the output circuit is isolated from the input circuit, and so the equivalent circuit is very similar, as illustrated in Fig. 4.12. The voltage equation for this case is

$$Ri + \frac{1}{C} \int i \, dt = 0. \tag{4.33}$$

The solution of this equation is

$$i = I_{max} \exp\left(-\frac{t}{CR}\right). \tag{4.34}$$

As in the discussion made for the step-down converter, by putting $t = T_{on}$, we can obtain the minimum current or the current just before the switch is again opened as follows:

$$I_{min} = I_{max} \exp\left(-\frac{T_{on}}{CR}\right) = I_{max}\left(1 - \frac{T_{on}}{CR}\right). \tag{4.35}$$

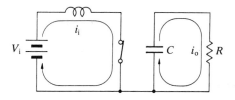

Fig. 4.12. In the T_{on} period the output and input circuits are separated from each other.

The peak-to-peak ripple is therefore expressed as

$$\Delta I = I_{max} - I_{min} = I_{max} \frac{T_{on}}{CR}. \tag{4.36}$$

Since, as before, I_{max} can be replaced with the average load current I_o as an approximation, the percentage ripple is given by

$$\text{Peak-to-peak ripple percentage} = 100 \frac{T_{on}}{CR}. \tag{4.37}$$

Exercise. Determine the minimum capacitance of the condenser to make the ripple percentage less than 5%, provided that the switching frequency is 5 kHz and the load resistance is 4 ohms.

Answers and explanations. As the maximum T_{on} time is 0.0002 s, we get

$$5 < 100 \times 0.0002/(4 \times C). \tag{4.38}$$

Hence

$$C > 10 - 5 = 10 \ \mu\text{F}. \tag{4.39}$$

Thus, the minimum capacitance is determined to be 10 μF.

4.3 Driving a direct-current motor with a step-down converter

It should be noted that when a d.c. motor is driven as the load of the d.c. converter of Fig. 4.5, the fundamental relation of eqn (4.4) does not hold in certain circumstances. We shall here discuss the necessary condition for this basic relation to hold, and develop a theory for determining the torque-versus-speed characteristics of a d.c. motor driven in the step-down PWM fashion.

4.3.1 Necessary condition for a continuous current

Let us assume that the d.c. motor is rotating at a constant speed, owing to its inertia, although pulsed voltage is applied to the motor in the PWM mode. As shown in Fig. 4.13(a), when the transistor is closed (ON), a current flows through the armature and the transistor as indicated by the solid curve. When the switch is opened, the current will circulate through the diode as depicted in (b). For the analysis of the current behaviour, let us discuss the problem of the equivalent circuit shown in Fig. 4.14, where the brush characteristics and also the transistor voltage drop are represented by two equivalent diodes D_B and D_T respectively.

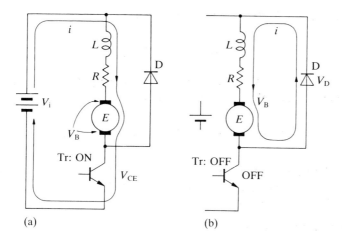

Fig. 4.13. When a d.c. motor is driven as the load of a step-down converter, the back e.m.f. in the armature works as a battery whose voltage E is proportional to the motor's speed. (a) Transistor ON; (b) transistor OFF.

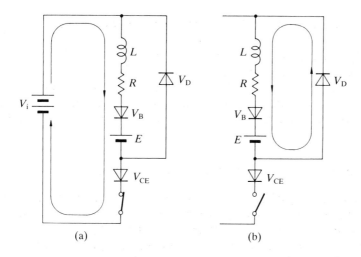

Fig. 4.14. Equivalent circuit for driving a d.c. motor; the brush characteristics are represented by a diode whose forward voltage is V_B, and the transistor's collector-to-emitter voltage drop V_{CE} by another diode. (a) Transistor closed; (b) transistor opened.

The current for the T_{on} period is governed by

$$L\frac{di}{dt} + Ri + E + V_B + V_{CE} = V_i, \tag{4.40a}$$

or

$$L\frac{di}{dt} + Ri = V_i - E - V_B - V_{CE}, \tag{4.40b}$$

where it is assumed that the battery has no internal impedance, and where

L = inductance of both armature winding itself and reactor connected in series

R = armature resistance and reactor resistance

E = back e.m.f. generated in the armature winding

V_B = brush drop, and

V_{CE} = collector-to-emitter voltage drop in the ON transistor,

L/R = electrical time constant T_E.

Since the right-hand side of eqn (4.40b) is positive, it is obvious that the current governed by this equation increases with time.

The equation that governs the current in the OFF period is

$$L\frac{di}{dt} + Ri + E + V_B + V_D = 0 \tag{4.41a}$$

or

$$L\frac{di}{dt} + Ri = -(E + V_B + V_D). \tag{4.41b}$$

Since the right-hand side of this equation is negative, a decrease in current will occur. Hence, there is a possibility that the current will fall to zero before the switch closes again. It should be noted that $-E$ and $-V_B$ in the right-hand side do not appear if the load is a combination of L and R only. (cf. eqn (4.17)).

When the repetition cycle is so short that current does not fall to zero before one cycle ends, it will increase cycle by cycle and attain stationary behaviour after several cycles, as illustrated in Fig. 4.15(a). In contrast, when the PWM frequency is so low that the current falls to zero within a cycle, owing to the back e.m.f; stationary behaviour is then, repetition of the current profile occurring in the first cycle as shown in Fig. 4.15(b).

In the former case, where the switching frequency or the inductance is so high that the current is continuous, the motor voltage waveform is the same as the case of the LR load. Hence the fundamental rule of eqn (4.4) will hold.

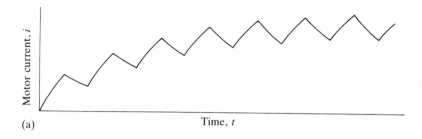

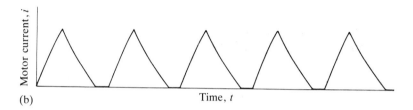

Fig. 4.15. Current waveforms in two cases when a d.c. motor is the load of a step-down converter. (a) When switching frequency or series inductance is high, the motor current is continuous and reaches the stationary state after a certain number of cycles. (b) When switching frequency or series inductance is not high enough, the current falls to zero before a switching cycle is completed.

Figure 4.16 illustrates how the armature voltage varies with time for the latter case. When the transistor is ON, the motor terminal voltage equals the input voltage V_i minus the transistor voltage drop V_{CE}. When the transistor is opened and the current circulates through the free-wheeling diode, the motor voltage is equal to the diode forward drop V_D. However, after the current stops, it will be E or the back e.m.f. Thus, this situation is similar to that of a large pulse-width ratio.

Let us study the conditions for the current to be continuous. Assuming that the current is zero just when the switch is closed, we obtain for the solution of eqn (4.40) the following:

$$i = \frac{A}{R}\left[1 - \exp\left(-\frac{R}{L}t\right)\right], \tag{4.42}$$

where

$$A = V_i - E - V_B - V_{CE}. \tag{4.43}$$

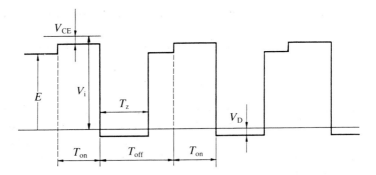

Fig. 4.16. Voltage waveforms appearing across the motor terminals when the current is discontinuous. After the current falls to zero before next cycle, the voltage is the back e.m.f. which results in an apparently higher time ratio.

Just before the switch opens, the current is

$$I_{END} = \frac{A}{R}\left[1 - \exp\left(-\frac{R}{L} T_{on}\right)\right]. \tag{4.44}$$

Next, when the switch is opened, the current is governed by

$$L \frac{di}{dt} + Ri = -B, \tag{4.45}$$

where

$$B = E + V_B + V_D. \tag{4.46}$$

The general solution of eqn (4.45) is

$$i = \frac{-B}{R} + Q \exp\left(-\frac{R}{L} t\right), \tag{4.47}$$

where Q is a constant to be determined from the initial condition that requires the current at $T = 0$ to be I_{max}, as was given by eqn (4.44). Hence we get

$$Q = I_{END} + \frac{B}{R}. \tag{4.48}$$

I_T, the current just before the switch is again closed, is given by

$$I_T = \left(I_{END} + \frac{B}{R}\right)\exp\left(-\frac{R}{L} T_{off}\right) - \frac{B}{R}. \tag{4.49}$$

If this value is negative, it means that the current falls to zero before the switch is closed again. On the other hand if the value is positive, it becomes the initial current for the following cycle, which means that the current increases cycle by cycle.

The time at which the current given by eqn (4.47) becomes zero is given by

$$T_Z = -\frac{L}{R} \log_e \left(\frac{B}{RQ}\right). \tag{4.50}$$

If $T_Z > T_{\text{off}}$, the current does not fall to zero, but is continuous. When this condition is satisfied, linear drooping characteristics will be seen in the torque vs. speed plots. However, if $T_Z < T_{\text{off}}$, the current falls to zero and hence the characteristics will not be linear. It is seen that this criterion is related to the following parameters:

(1) supply voltage (V_i);
(2) motor speed N, because back e.m.f. (E) is related to the speed as given by

$$E = K_E N, \tag{4.51}$$

where K_E is a constant known as the 'back e.m.f. constant';
(3) voltage drop in the free-wheeling diode (V_D);
(4) brush voltage drop (V_B);
(5) armature resistance (R); and
(6) the inductance in the armature winding and series inductor (L)

4.3.2 Computing average current

For determining the torque-versus-speed characteristics, calculation of the average current is required, as the motor torque T is related to the current by

$$T = K_T i, \tag{4.52}$$

where K_T is the torque constant and this is theoretically identical to the back e.m.f. constant.

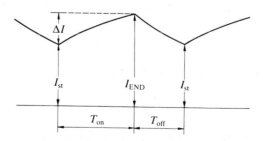

Fig. 4.17. Stationary-state waveforms for the case of continuous current.

The average current in the stationary state can be computed as followings, using Fig. 4.17. We assume that in the ON period the current increases from I_{st}, and get the solution of eqn (4.40) as

$$i = \frac{A}{R} - \left(\frac{A}{R} - I_{st}\right) \exp\left(-\frac{t}{T_E}\right), \tag{4.53}$$

where A is given by eqn (4.43). I_{max}, or the current just before the switch is closed, is therefore given by

$$I_{END} = \frac{A}{R} - \left(\frac{A}{R} - I_{st}\right) \exp\left(-\frac{T_{on}}{T_E}\right). \tag{4.54}$$

In the OFF period, the current is given by eqn (4.47). However, it should be noted that the constant Q appearing in eqn (4.47) must be different from the Q that is given by eqn (4.48), because now the starting value is not zero but I_{st}. The expression for Q for this case is

$$Q = \frac{A + B}{R} - \left(\frac{A}{R} - I_{st}\right) \exp\left(-\frac{T_{on}}{T_E}\right). \tag{4.55}$$

Since the current is continuous, the value when the switch is again closed must be the same as I_{st}:

$$I_{st} = -\frac{B}{R} + Q \exp\left(-\frac{T_{off}}{T_E}\right). \tag{4.56}$$

substituting Q of eqn (4.55) into this equation, we get

$$\left[1 - \exp\left(-\frac{1}{T_E f}\right)\right] I_{st} = \\ -\frac{B}{R} + \frac{A + B}{R} \exp\left(-\frac{T_{off}}{T_E}\right) - \frac{A}{R} \exp\left(-\frac{1}{T_E f}\right), \tag{4.57}$$

where the following relation has been used:

$$T_{on} + T_{off} = \frac{1}{f}. \tag{4.58}$$

From eqn (4.57), I_{st} is derived as

$$I_{st} = \frac{(A + B) \exp(-T_{off}/T_E) - A \exp(-1/(T_E f)) - B}{R[1 - \exp(-1/(f T_E))]}. \tag{4.59}$$

We shall continue discussion of Fig. 4.16 to get the mean current. Considering eqn (4.53), the integrate S_{on} of the current in the ON period is given by

$$S_{on} = \int_0^{T_{on}} \left[\frac{A}{R} - \left(\frac{A}{R} - I_{st}\right) \exp\left(-\frac{t}{T_E}\right)\right] dt \tag{4.60}$$

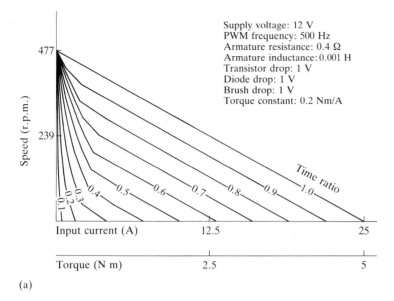

(a)

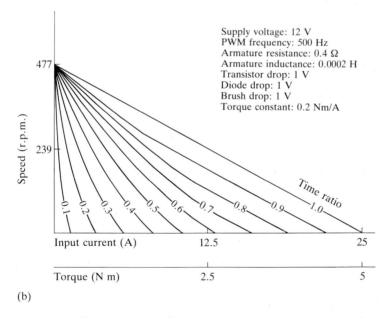

(b)

Fig. 4.18. Computed speed-versus-current/torque characteristics of a d.c. motor diven by a step-down converter. (a) Switching frequency = 500 Hz, armature inductance = 1 mH. (b) Switching frequency = 500 Hz, armature inductance = 0.2 mH.

$$= \frac{A}{R} T_{on} - \left(I_{st} - \frac{A}{R}\right) T_E \left[1 - \exp\left(-\frac{t}{T_E}\right)\right]. \qquad (4.61)$$

Likewise, using eqn (4.47), the integral S_{off} of the current in the OFF period is

$$S_{off} = \int_0^{T_{off} \text{ or } T_Z} \left[\frac{B}{R} + Q \exp\left(-\frac{t}{T_E}\right)\right] dt \qquad (4.62)$$

It should be noted that when the current falls to zero before T_{off}, the integration should be performed from 0 to T_Z which is given by eqn (4.50).

The average current is hence given by

$$\langle I \rangle = (S_{on} + S_{off})f. \qquad (4.63)$$

4.3.3 Torque-versus-speed characteristics

The relation between the torque and speed can be computed by determining the relation between the current and the back e.m.f., E, since the torque is proportional to the current as given by eqn (4.52) and the speed to the back e.m.f. as given by eqn (4.51).

Figure 4.18 shows two examples calculated using the above theory. The speed-versus-torque relation displays linear, parallel decreasing characteristics in the continuous-current region and curved characteristics in the discontinuous region. It is seen that with increasing switching frequency the linear region dominates. Table 4.1 is a BASIC program listing for computing/drawing the characteristic curves. The flowchart for drawing these curves is given in Fig. 4.19.

4.3.4 Current ripple when the motor is driven

It can be shown that the ripple percentage when the motor is the load is given by an expression very similar to eqn (4.29).

The peak-to-peak ΔI seen on the increasing curve of the current in Fig. 4.17 is

$$\Delta I = \left(\frac{A}{R} - I_{st}\right)\left[1 - \exp\left(-\frac{T_{on}}{T_E}\right)\right]. \qquad (4.64)$$

The approximation of I_{st} can be derived from eqn (4.59) by substituting the following approximations:

$$\exp\left(-\frac{T_{off}}{T_E}\right) = 1 - \frac{T_{off}}{T_E} \qquad (4.65)$$

and

$$\exp\left(-\frac{1}{T_E f}\right) = 1 - \frac{1}{T_E f}. \tag{4.66}$$

The result is

$$I_{st} = \frac{f}{R}(A T_{on} - B T_{off}) \tag{4.67}$$

or

$$I_{st} = \frac{A T_{on} - B T_{off}}{R(T_{on} + T_{off})}. \tag{4.68}$$

A similar approximation can be applied to eqn (4.64) and we get

$$\Delta I = \left(\frac{A}{R} - I_{st}\right)\frac{T_{on}}{T_E}. \tag{4.69}$$

Substituting eqn (4.68) into eqn (4.69) and considering $T_E = L/R$, we get

$$\Delta I = \frac{A + B}{L}\frac{T_{on} T_{off}}{T_{on} + T_{off}}. \tag{4.70}$$

Here $A + B$ is composed as

$$A + B = (V - E - V_B - V_{CE}) + (E + V_B + V_D) \tag{4.71}$$

$$= V + V_D - V_{CE}. \tag{4.72}$$

Therefore,

$$\Delta I = \frac{V + V_D - V_{CE}}{L}\frac{T_{on} T_{off}}{T_{on} + T_{off}}. \tag{4.73}$$

If the diode forward voltage drop and transistor drop V_{CE} are almost the same, eqn (4.73) is the same as eqn (4.29), which was derived for an LR load.

4.4 Examples and applications of direct-current converters

(1) *Simple step-down converter.* Figure 4.20 shows a circuit diagram of a very simple step-down d.c. converter. The PWM signal is generated by an astable multivibrator consisting three transistors (Tr_1 to Tr_3), and the switch element for the power circuit is a Darlington connection of two transistors Tr_4 and Tr_5. The astable multivibrator of this circuit provides a fixed-frequency control of about 1 kHz, and the pulse-width ratio is adjusted by varying the ratio of R_1 and R_2 using a rheostat of 10 kΩ.

Table 4.1. Program listing for computing the speed-versus-torque characteristics of a d.c. motor driven by a step-down converter.

```
1000 REM SPEED-VERSUS-TORQUE/CURRENT CHARACTERISITCS
1010 CLS
1020 '               **** PARAMETERS ****
1030 V=12
1040 RA=.4
1050 LA=.0004
1060 K=.2
1070 VD=1.0
1080 VB=1.0
1090 VCE=1.0
1100 FREQ=500
1110 '               **** LABELLING ****
1120 LOCATE 45,2,1:  PRINT "Supply voltage:";      V;"V"
1130 LOCATE 45,3,1:  PRINT "PWM frequency:";       FREQ;"Hz"
1140 LOCATE 45,4,1:  PRINT "Armature resistance:";RA;"ohms"
1150 LOCATE 45,5,1:  PRINT "Armature inductance:";LA;"H"
1160 LOCATE 45,6,1:  PRINT "Transistor drop:";     VCE;"V"
1170 LOCATE 45,7,1:  PRINT "Diode drop:";          VD;"V"
1180 LOCATE 45,8,1:  PRINT "Brush drop:";          VB;"V"
1190 LOCATE 45,9,1:  PRINT "Torque constant:";     K;"Nm/A"
1200 LOCATE 58,15,1: PRINT "Time ratio"
1210 LOCATE 0,15,1:  PRINT "Speed"
1220 LOCATE 0,16,1:  PRINT "(rpm)"
1230 LOCATE 5,19,1:  PRINT "Input current (A)"
1240 LOCATE 5,22,1:  PRINT "Torque (Nm)"
1250 LOCATE 15,24,1
1260 PRINT "Speed-versus-current/torque characteristics"
1270 '
1280 '     **** DRAWING OF AXES/SCALES ****
1290 LINE (50,280)-(50,10),    PSET
1300 LINE (50,280)-(570,280), PSET
1310 LINE (50,320)-(570,320), PSET
1320 LINE (550,280)-(550,285),PSET
1330 LINE (300,280)-(300,285),PSET
1340 LINE (300,320)-(300,325),PSET
1350 LINE (550,320)-(550,325),PSET
1360 IMAX=(V-VB-VCE)/RA
1370 S$=STR$(IMAX)
1380 T$=STR$(IMAX/2)
1390 U$=STR$(IMAX*K)
1400 V$=STR$(IMAX*K/2)
1410 SYMBOL (530,290),S$
1420 SYMBOL (280,290),T$
1430 SYMBOL (530,330),U$
1440 SYMBOL (280,330),V$
```

Table 4.1. *continued*

```
1450  Y=(V-VB-VCE)*30/K/3.1416
1460  W$=STR$(INT(Y+.5))
1470  Z$=STR$(INT(Y/2+.5))
1480  D=280-20*(V-VB-VCE)
1490  SYMBOL (1,D-8),W$
1500  SYMBOL (1,(270+D)/2),Z$
1510  LINE (45,D)-(50,D),PSET
1520  LINE (45,140+D/2)-(50,140+D/2),PSET
1530  '
1540  '     **** CALCULATION/GRAPH-DRAWING ****
1550  TE=LA/RA
1560  TC=1/FREQ
1570  C=INT(V-VB-VCE)
1580  DIM F(C+2,10)
1590     FOR TP=1 TO 10
1600     TR=.1*TP
1610     TON=TC*TR
1620     TOF=TC*(1-TR)
1630        FOR E=0 TO C+1
1640        A=V-E-VCE-VB: B=E+VD+VB+.0001
1650        IEND=A/RA*(1-EXP(-TON/TE))
1660        ALPHA=IEND+B/RA
1670        TZ=-TE*LOG(B/RA/ALPHA)
1680        IST=0
1690        IF TZ>TOF THEN GOSUB 1860
1700        SON=TON*A/RA+(IST-A/RA)*TE*(1-EXP(-TON/TE))
1710        SOF=-(TE*ALPHA*(EXP(-TZ/TE)-1))-TZ*B/RA
1720        IAV=(SON+SOF)*FREQ
1730        F(E+1,TP)=50+500*IAV/IMAX
1740        NEXT E
1750     NEXT TP
1760     FOR TR=1 TO 10
1770     T$=STR$(.1*TR)
1780     SYMBOL (F(2,TR)-5,245),T$
1790        FOR E=1 TO C
1800        LINE (F(E,TR),280-20*(E-1))-(F(E+1,TR),280-20*E),PSET
1810        NEXT E
1820     NEXT TR
1830  END' OF CALCULATION/DRAWING
1840  '
1850  '            **** SUBROUTINE ****
1860  TZ=TOF
1870  G=-1/TE
1880  IST=((A+B)*EXP(-TOF/TE)-A*EXP(G/FREQ)-B)/RA/(1-EXP(G/FREQ))
1890  ALPHA=A/RA+(IST-A/RA)*EXP(-TON/TE)+B/RA
1900  RETURN
```

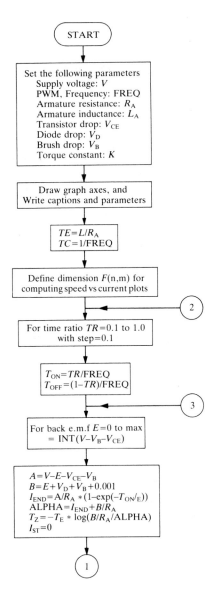

Fig. 4.19. Flowchart for computing the current/torque-versus-speed characteristics of a d.c. motor driven by a step-down d.c. converter.

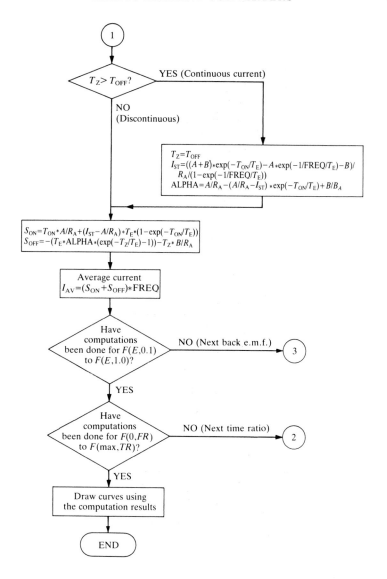

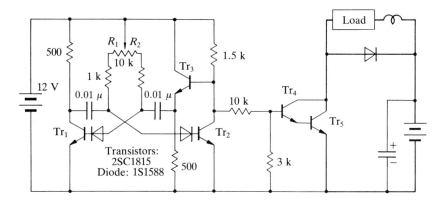

Fig. 4.20. Step-down converter using an astable multivibrator as the PWM signal generator.

(2) *Simple step-up converter.*   Figure 4.21 is a step-up converter that converts an input of 12 V to a voltage from 13 V up to about 40 V with a switching frequency of about 1 kHz. The same astable multivibrator as before is also used here.

(3) *Power controller for a laboratory-made electric vehicle.*   The circuit diagram of Fig. 4.22 is for a d.c. converter for controlling the voltage applied to a motor used in an electric vehicle that was designed in the author's laboratory as a students' final-year project. Figure 4.22(a) is a block diagram explaining the function of each part; part (b) shows the details of the power

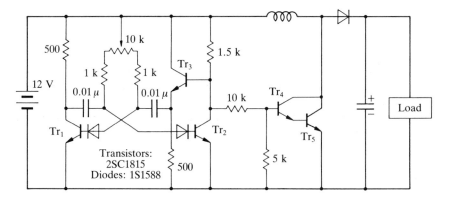

Fig. 4.21. Step-up converter using an astable multivibrator as the PWM signal generator.

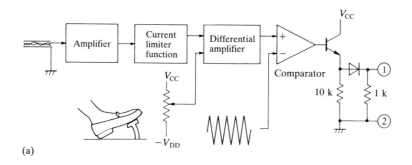

(a)

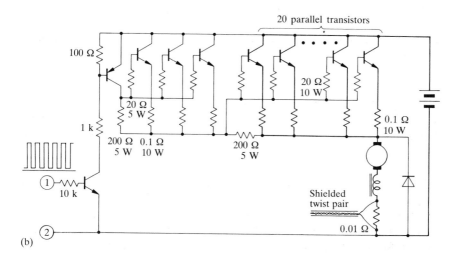

(b)

Fig. 4.22. Control portion (a) and power stage circuit (b) of a step-down converter designed for the drive on an electric vehicle motor.

circuit. The d.c. motor used in this project is a series-wound type having a shunt field for safety purposes, but a permanent-magnet motor is also suitable. The 20 parallel-connected transistors are switched at 2 kHz to control a maximum current of 70 A. The acceleration pedal is connected to a rheostat to adjust the pulse-width ratio.

References

1. Bedford, B.D. and Hoft, R.G. (eds.) (1964). *Principles of inverter circuits*, Chapter 10. Wiley, New York.

5 Servo-amplifiers

The power stage of the electronic circuit designed for driving a servomotor via speed/position controls is referred to as a servo-amplifier, and is extensively employed in industrial robots and other numerically controlled equipment. The fundamental type of servo-amplifier is the type that uses transistors in the linear region, and the pulse-width modulated scheme is the favoured energy-saving strategy. This chapter focuses first on the principles of several types of servo-amplifiers for conventional d.c. motors, and our discussion will be extended to elucidate the relation between the linear and PWM means. Some practical circuit diagrams are also presented. Servo-amplifiers for driving a.c. or brushless d.c. servomotors are dealt with in Chapter 8.

5.1 Linear and PWM servo-amplifiers

Certain types of servo-amplifiers are dealt with in Chapter 8 of Reference [1]. A different approach will be employed here, better fitted to the philosophy of the ideas developed in this book.

Servo-amplifiers are classified into two basic categories according to the method they employ to drive soild-state devices. One, which is referred to as the linear servo-amplifier, drives bipolar transistors in their linear or active regions; the other is the PWM servo-amplifier, which drives bipolar transistor or MOSFETs in the switching mode, as employed in a d.c. converter.

We shall start with the linear servo-amplifier, since this is simpler than the PWM one in circuit construction. The elementary linear servo-amplifier is as shown in Fig. 5.1. A d.c. motor forms the load of an emitter follower circuit, which means that one terminal of the motor is connected to the transistor's emitter and the other to the GND or negative terminal of the power supply. The control signal is applied to the transistor base with respect to the GND. If we ignore the base–emitter forward potential, which is about 0.6 V, the control or input voltage v_i appears across the motor terminal. The motor current is, however, supplied from the power source E. Thus the motor is controlled by the input voltage.

This elementary circuit has an essential drawback. That is, the potential applied to the motor is unipolar, which means that the motor is driven in only one direction. As servomotors are often used for position control in which bidirectional motion is essential, it is required that both polarity and magni-

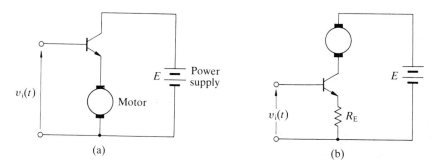

Fig. 5.1. Two basic drive circuits of d.c. motors: (a) voltage-control type, (b) current-control type.

tude of the voltage applied to the motor terminals are controlled by the input signal. Figure 5.2 shows the basic principle of the servo-amplifier to drive a motor in either direction. This circuit uses a PNP transistor and an NPN transistor in a complementary emitter-follower scheme.

If we ignore the base–emitter forward potentials in both transistors, the input voltage emerges at the emitter of each transistor, and this is applied to the motor. In this circuit the motor current flows through either Tr_1 or Tr_2, and the transistors also operate in the linear region. For example, when the voltage v_i is positive, Tr_1 carries a current that is supplied from the power supply E_1. Since the PN junction in this transistor is forward-biased, a voltage of approximately 0.6 V appears across the base and emitter. At this time, however, the base–emitter PN junction in Tr_2 is reverse-biased and this transistor is in the cutoff region. Conversely, when v_i is negative, Tr_2 closes and carries the motor current, which is supplied by the power source E_2.

Another problem emerges with this circuit. When we consider the base–emitter forward potential of each transistor, the relation between v_i and v_o becomes as shown in Fig. 5.3(a); a dead zone (or backlash) of about −0.6 to

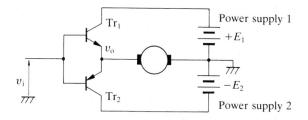

Fig. 5.2. A basic bipolar servo-amplifier of the voltage-control type.

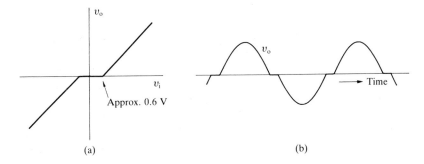

(a)   (b)

Fig. 5.3. Characteristics of the servo-amplifier in Fig. 5.2: (a) v_o versus v_i relations, and (b) waveform of v_o with respect to time.

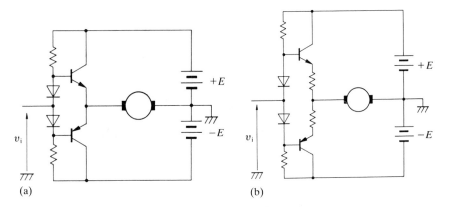

(a)   (b)

Fig. 5.4. Eliminating dead zones by employing two rectifier diodes. (a) Forward voltage across the diode compensates the forward voltage in the base-to-emitter PN junction, but (b) when compensation is incomplete, two low resistors are placed between the emitters of the transistors.

$+0.6$ V appears. Hence, when v_i varies as a sine wave, the waveform of v_o is distorted as shown in Fig. 5.3(b). In Fig. 5.4(a) this defect is removed by adding two rectifier diodes, and the influence of the forward-biased base–emitter junction in both transistors is compensated for by the PN junctions of the two diodes. When this compensation is incomplete, two resistors of low resistance are put between both emitters as shown in Fig. 5.4(b) to make up the small difference between the junction characteristics in the transistors and the diodes.

5.2 Servo-amplifiers having a voltage gain

In the servo-amplifiers shown so far, the input voltage v_i emerges at the motor
terminals, which means that the input voltage signal is not amplified before it
is applied to the motor. Some servo-amplifiers function to amplify the input
signal and apply it to the motor. Figure 5.5 shows the principle of this type of
servo-amplifier for unidirectional drive. Three current paths are shown, and
the magnitude of the current is indicated by the thickness of the curves. If the
base–emitter forward potential in Tr_1 is negligible, the voltage appearing at
point A, or the voltage across R_A, is the same as the input voltage

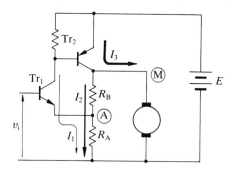

Fig. 5.5. Voltage-controlled servo-amplifier having gain $(R_A + R_B)/R_A$.

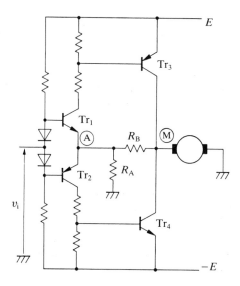

Fig. 5.6. A bipolar-driven amplifier having voltage $(R_A + R_B)/R_A$.

v_i. Note that I_2 is much bigger than I_1, owing to the current amplification in Tr$_2$. Hence we can ignore the effect of I_1 in the computation of the voltage that appears at point M to be applied to the motor. It is v_i multiplied by $(R_A + R_B)/R_A$.

A bidirectional servo-amplifier developed from this idea is shown in Fig. 5.6. Two diodes are also used for compensation of the influence of the forward-biased P N junctions in Tr$_1$ and Tr$_2$.

5.3 Servo-amplifier of the current-control type

In the servo-amplifiers presented so far, the motor voltage is controlled by the input signal. In these types of amplifiers the motor current is determined not only by the voltage but also by the motor parameters and the speed, which may cause a motional delay in response to the input signal in feedback controls and result in an unstable motion. In many applications, therefore, it is preferred that the motor current is directly controlled by the input signal, because the torque developed in the motor, which is the most important factor for creating the rotational motion in speed/position controls, is directly proportional to the motor current as given by

$$m = K_T i_m \qquad (5.1)$$

where m = instantaneous torque
$\quad\quad\ i_m$ = motor current
$\quad\quad\ K_T$ = torque constant of the motor.

Hence, many practical servo-amplifiers are designed in such a way that the input signal controls the motor current rather than the voltage. Current control is available by adding a current detector and an operational amplifier, for example as shown in Fig. 5.7. Here a low resistance R_s is used as the current detector. The operational amplifier used in this scheme works so as to make the potential at the negative terminal the GND potential. Hence, the relation between v_s (the terminal voltage of R_s) and v_i (the input voltage) is

$$v_s = -\frac{R_2}{R_1} v_i, \qquad (5.2)$$

where $R_s i = v_s$.

Therefore the motor current is determined by the input voltage v_i:

$$i_m = -\frac{R_2}{R_1 R_s} v_i. \qquad (5.3)$$

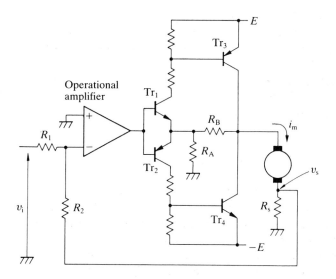

Fig. 5.7. A bidirectional amplifier of the current-controlled type.

5.4 PWM servo-amplifiers

One feature of linear servo-amplifiers is that their circuits are simple and do not generate harmful electrical noise. However, a lot of power is dissipated as heat in the final-stage transistors and therefore a large heat sink is needed to remove the heat and protect the transistors from thermal damage. For example, as shown in Fig. 5.8(a), when a voltage of 20 V is applied between the emitter and collector of a power transistor, and it carries a collector current of 3 A, the power loss in it is 3 A × 20 V = 60 W.

To reduce the heat loss in transistors and improve the efficiency of servo-amplifiers, the pulse-width modulation technique should be employed as in d.c. converters discussed in Chapter 4. In a PWM amplifier, a transistor is either in the saturation region or in the cutoff region (see Fig. 5.8(b)). When the transistor is in the saturation region or the fully ON state, the potential between collector and emitter is about 1 V or a little higher, which means that the power dissipation is reduced. When the transistor is in the cutoff region or the fully OFF state, it carries negligible current, and therefore the power loss is correspondingly negligible.

However, when the ON and OFF states are repeated in a high-frequency switching mode, the power loss may not be negligible because the transistor experiences, many times per second, a transient linear region that exists

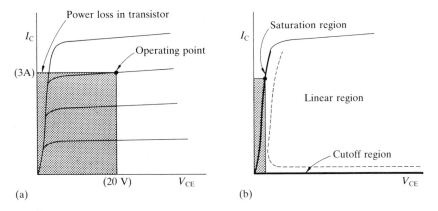

Fig. 5.8. Power loss inside a transistor: (a) linear servo-amplifier dissipates much electrical power into heat, while (b) P W M amplifier produces less heat loss.

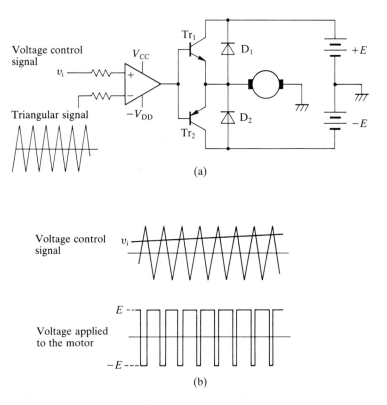

Fig. 5.9. A basic servo-amplifier of the voltage-controlled type: (a) circuit and (b) waveforms.

between saturation and cutoff. Hence, quick switching is essential in designing a high-efficiency P W M servo-amplifier.

5.4.1 *Principle of voltage-control P W M amplifiers*

Figure 5.9 illustrates the basic principle of the P W M amplifier of the voltage-
controlled type, which is similar to the linear servo-amplifier shown in
Fig. 5.4. A major difference in the circuit construction is that a rectifier diode
is connected in parallel with each transistor. The function of these diodes was
discussed in Chapters 2 and 4. Another addition is a comparator I C, which is
a type of generator for the pulse-width modulated signal, or a modulator,
and is put prior to the power circuit. The comparator has two input terminals.
The voltage control signal is applied to the (+) terminal, and a triangular
signal is applied to the (−) terminal. The relationship between the input
signal, the output signal on the comparator, and the motor potential are as
follows.

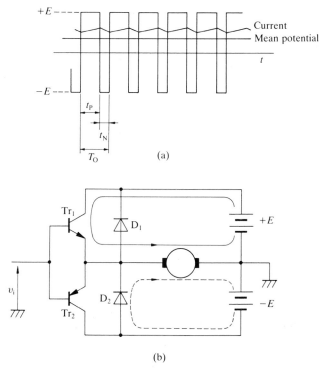

(a)

(b)

Fig. 5.10. (a) The pulse-width modulated voltage and current waveforms, and (b) the
current paths. As the mean voltage is now positive, current flows through Tr_1 when it
is ON and Tr_2 is OFF, while current flows through D_2 when Tr_1 is OFF and Tr_2 is
ON.

(1) When v_i is greater than the triangular signal, the output voltage is always equal to $+V_{CC}$, and transistor Tr_1 is brought into the fully ON state (saturation region), while Tr_2 is in the fully OFF state (cutoff region). Therefore the potential applied to the motor is E.

(2) When v_i is less than the triangular signal, the output voltage is always $-V_{DD}$, and transistor Tr_2 is brought into the fully ON state, while Tr_1 is in the fully OFF state. Therefore the potential applied to the motor is $-E$.

The relation between the potential waveform across the motor and its mean voltage is explained in Fig. 5.10(a). In this figure, the period t_P in which Tr_1 is ON is longer than t_N in which Tr_2 is ON. Hence the current flows from left to right in the motor as shown by the solid path in Fig. 5.10(b). Since this current is supplied by the $+E$ source, the current polarity is defined as positive. In this period the current increases with time. The current path for the t_N period is shown by the broken path in the same figure. The current flows through diode D_2 and this is fed back to the power supply $(-E)$. Note that this current decreases with time. When the switching frequency is high, the next t_P period comes before the current falls to zero. As discussed in Section 4.2, the higher the frequency, the lower the ripple component in the motor current. In some of the recent PWM servo-amplifiers, the pulse-width modulation is implemented at a frequency as high as 20 kHz or higher so as not to generate harmful audio noise.

When an ideal, loss-free pulse-width modulation is implemented, the mean output voltage of the PWM servo-amplifier will be a linear function of the time ratio t_P/T_O, as shown in Fig. 5.11.

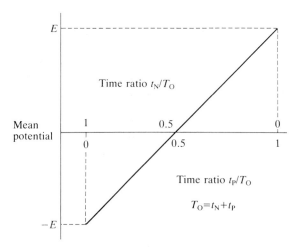

Fig. 5.11. Output voltage of the PWM amplifier shown in the Fig. 5.10 as a function of the time ratio.

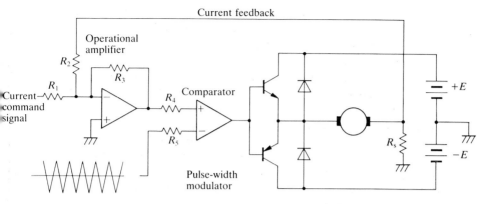

Fig. 5.12. A PWM servo-amplifier of the current-controlled type.

5.4.2 *Current-controlled PWM amplifiers*

Figure 5.12 is the current-controlled scheme for a PWM servo-amplifier. Prior to the pulse-width modulator, an operational amplifier is used to control the motor current using a current sensor R_s as previously explained in relation to Fig. 5.7.

5.5 PWM servo-amplifiers in the bridge scheme

The servo-amplifiers so far presented require two d.c. power sources. We may eliminate one power supply by employing a bridge circuit. Bridge schemes are available for both the linear and PWM servo-amplifiers. However, explanation will be focused on the PWM arrangements.

As shown in Fig. 5.13, in the fundamental scheme of the bridge amplifier, four transistors are used to control the voltage waveform applied to a d.c.

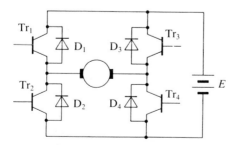

Fig. 5.13. Fundamental scheme of the H-bridge servo-amplifier.

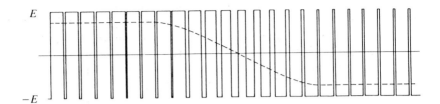

Fig. 5.14. PWM waveform that swings between E and $-E$ using four transistors in the switching mode.

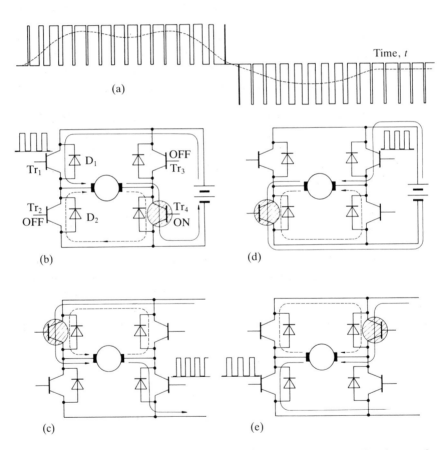

Fig. 5.15. (a) PWM waveform that swings between E and zero or $-E$ and zero using two transistors in the switching mode. (b) Tr_4 is ON and Tr_1 is in PWM mode; (c) Tr_1 is ON and Tr_4 is in PWM mode; (d) Tr_2 is ON and Tr_3 is in PWM mode; (e) Tr_3 is ON and Tr_2 is in PWM mode.

servomotor. There are two basic ways of using the four transistors. One, which always uses these transistors in the switching mode, yields voltage waveforms as shown in Fig. 5.14, in which the instantaneous voltage swings between $+E$ and $-E$ and the average voltage is indicated by the curve. In this method, when Tr_1 and Tr_4 are ON Tr_2 and Tr_3 are OFF, $+E$ is applied to the motor and, conversely, $-E$ emerges across the motor when Tr_1 and Tr_4 are OFF and Tr_2 and Tr_3 are ON. When the positive period t_P equals the negative period t_N, the time-average voltage is zero.

The other method generates waveforms as shown in Fig. 5.15(a). Use of the four transistors is shown in (b) to (e); when a positive potential is to be applied to the motor, Tr_4 is kept ON and Tr_2 and Tr_3 are OFF, while Tr_1 is driven in the switching (PWM) mode as in (a). Let us see more in detail how the motor current flows in this scheme. When Tr_1 is ON, the current path will be the solid path shown in the figure and the supply voltage is applied to the motor if the diode forward voltages are ignored. On the other hand, when Tr_1 is OFF, so that only Tr_4 is ON, the motor current will take the broken path, and current circulates through the motor, Tr_4, and D_2 owing to the inductive effect of the motor.

Figure 5.15(c) shows another scheme for using the transistors. Tr_1 is kept always ON and Tr_4 is used in the switching mode. Figures 5.15(d) and (e) show the case in which a negative voltage is to be applied to the motor.

Let us compare the merits and demerits of these two methods. In the former, the voltage transition across zero is carried out smoothly in either direction—from negative to positive or vice versa. Such a smooth transition is a very important requirement in a servo-amplifier. In contrast, the second method may cause an abrupt transition across zero owing to small differences in switching characteristics of each transistor used.

The feature of the latter method is that ripple percentage in the motor current is half that in the former at the same frequency and average current.

It is, naturally, possible to combine these two methods to produce a waveform as shown in Fig. 5.16. In Section 7.9 a rather more sophisticated method will be presented in discussion of the application of pulse-width modulation to inverters.

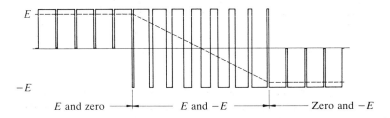

Fig. 5.16. Combination of the two methods illustrated in Figs. 5.14 and 5.15.

5.6 Further considerations of the PWM technique

Pulse-width modulation is a method of expressing analogue signals in the form of one-bit time-varying signals. We consider further the generation and demodulation of PWM signals.

5.6.1 *Analogue, hybrid, and digital methods of generating PWM signals*

In the explanation of the generation of pulse-width modulated waveforms in Section 5.4.1 using Fig. 5.9, a comparator IC was used. This is again illustrated in Fig. 5.17(a); a triangular wave is applied to one of the two terminals

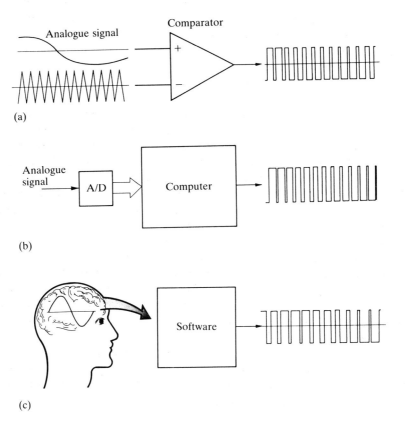

Fig. 5.17. Methods of generating PWM signals. (a) Using a comparator, the analogue signal is modulated by a modulation wave. (b) Using an analogue-to-digital converter, the analogue signal is converted into digital data to be fed to a computer that generates the PWM signal. (c) An entirely digital method: the analogue quantity is in the engineer's mind.

of the comparator to modulate the analogue signal that is applied to the other terminal. The PWM waveform is sent out from the output terminal. Now let us define the analogue signal to be modulated the source signal, and the triangular signal as the modulation wave.

Instead of using a comparator IC, which is a sort of analogue device, a set of digital devices or a microprocessor can be used as shown in (b). The source signal is converted into digital data to be supplied to the microcomputer, and the data are dealt with by a program whose function is similar to that of comparator.

Furthermore, it is possible to use no analogue signal in some cases, as shown in (c). The source signal has been generated in the programmer's or designer's mind, and the function of producing PWM signals has been assembled in the microcomputer's memory area. This is a completely digital method. Even in this case we can consider the analogue signal in the programmer's mind as existing as a real signal.

5.6.2 *Unipolar, bipolar, and H-bridge schemes*

Figure 5.18 shows three different PWM methods of driving a motor. Though these have already been dealt with in this chapter, we survey them here to consider the bias voltages applied to the comparator and the power circuit voltages. The first is the unipolar or unidirectional scheme, which can be employed in a power circuit to drive a d.c. motor only in one direction, as shown in (a). When the circuit is a common-collector type, the signal voltage is not restricted by the voltage of the drive circuitry. However, if an emitter follower is employed, the high level of the signal voltage must be the same as the battery voltage E in the drive circuit. The modulation signal must swing from 0 to E, and the source signal must also vary within this range.

The second method is the bipolar or bidirectional scheme for driving in either direction. Usually, as shown in Fig. 5.18(b), a bipolar emitter follower is employed. Hence, the bias potential applied to the comparator is $+E$ and $-E$; the modulation signal swings between these two levels, and the source signal can vary within this range.

The third scheme uses the H-bridge circuit to drive the machine, again in bidirectional fashion but from two channels of the unipolar PWM signals. The PWM wave applied to the motor is the difference between the two channel signals, as shown in Fig. 5.14 to 16.

5.6.3 *Demodulation of PWM signals*

To discuss the meaning of demodulation, which literally means recovering an analogue signal from the pulse signals, let us take a d.c. motor as an example.

The equivalent circuit of a d.c. motor is as shown in Fig. 5.19(a) (see

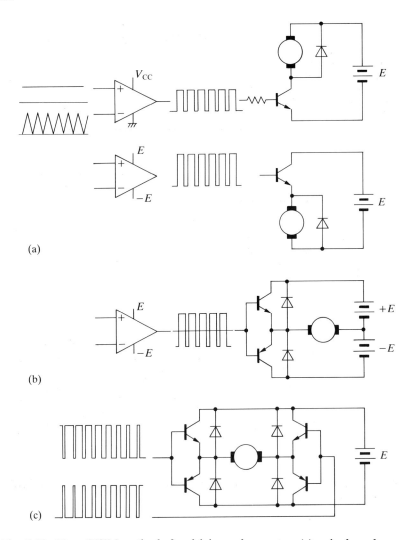

Fig. 5.18. Three PWM methods for driving a d.c. motor: (a) unipolar schemes of common-collector and emitter-follower types; (b) bidirectional scheme using two power supplies; (c) H-bridge for bi-directional drive using one power supply.

Chapter 7 of Reference [1].). The principle of demodulation is best illustrated when the motor is at standstill. Here the capacitor, which represents the rotor, is regarded as being short-circuited, i.e. R_D, representing the mechanical output power, is regarded as being 0. Hence, the circuit is simply as in (b). The relation between the pulse-width-modulated voltage and the current is as shown in (c).

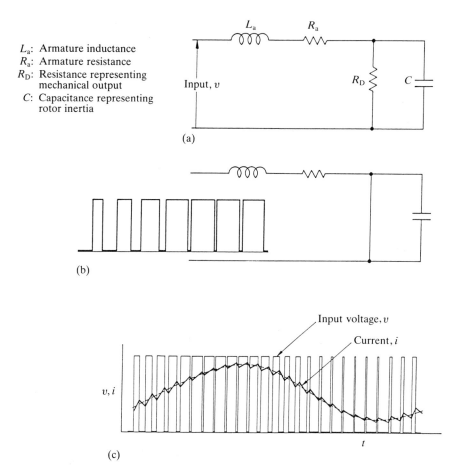

L_a: Armature inductance
R_a: Armature resistance
R_D: Resistance representing
      mechanical output
C: Capacitance representing
      rotor inertia

Fig. 5.19. Equivalent circuit of a d.c. motor: (a) general circuit; (b) for standstill; (c) current waveform.

If the ripple in the current is not taken into consideration, and the current is supposed to vary along the central smooth curve, then this current is the same as the current that would flow if the analogue source were applied to the d.c. motor. The higher the modulation frequency, the less the ripple component in the current. To describe this more quantitatively, when the pulse period is much shorter than the electric time constant (L_a/R_a in this case), the ripple component is negligible. If a smooth movement is required, the pulse period should be selected to be much shorter than the mechanical time constant of the motor, which is JR_a/K^2.

In this model it is not necessary to recover the source voltage signal, because in a motor the pulse-width modulated voltage can be considered to be demodulated in the form of current, which is the essential quantity for producing torque.

5.6.4 *First-order passive filter*

In many instances it *is* required to demodulate the PWM signals to observe the source voltage waveform, for example when the completely digital method presented in Fig. 5.17(c) is employed. The simplest method is a CR low-pass filter as shown in Fig. 5.20(a). An example of an original signal and its pulse-width modulated signal is seen in (b). Figure (c) is the demodulated waveform appearing across the capacitor.

As seen from the Bode diagram in (d), a drawback of this method is that if the constant CR is large enough to eliminate the ripple components, a considerable time-lag may be seen in some frequency components. For example, the fundamental-wave frequency is 50 Hz, and the cutoff frequency f_c is set at 10 times as high as this. Now the time constant CR is $1/(2\pi f_c) = 3.18$ ms. If the carrier-pulse frequency is 5 kHz, which is again 10 times as high as the cutoff frequency, the gain for the carrier is -20 dB; that is, not low enough.

On the other hand, the time-lag seen in the demodulated fundamental-frequency component is as small as $6°$. However, for the 5th harmonic component it is as much as $27°$. If the time constant is increased to 10 ms, the gain for the carrier is improved to be about -30 dB, but the phase delay for the 5th harmonics is further degraded to $56°$. This means that the original signal is recovered in a very different waveform.

5.6.5. *Second-order active filter*

For certain cases a second-order active filter, shown in Fig. 5.21(a) or (b), is recommended because, as indicated in its Bode diagram shown in (c), by selecting suitable capacitors and resistors used in the filter, the time lag in the demodulation of the original component can be made small and the ripple component of the carrier can be frequency reduced to a negligible level.

The cutoff frequency f_c and the damping factor ζ are given respectively by

$$f_c = \frac{1}{2\pi(C_1 C_2 R_1 R_2)^{\frac{1}{2}}} , \tag{5.4}$$

$$\zeta = \frac{C_2(R_1 + R_2)}{2(C_1 C_2 R_1 R_2)^{\frac{1}{2}}} . \tag{5.5}$$

As an example, if we choose the parameters as $R_1 = 1.0$ kΩ, $R_2 = 10$ kΩ, $C_1 = 0.68$ μF, and $C_2 = 0.015$ μF, the cutoff frequency is about 500 Hz and the damping factor is 0.258. Using this filter, frequencies up to 250 Hz can be

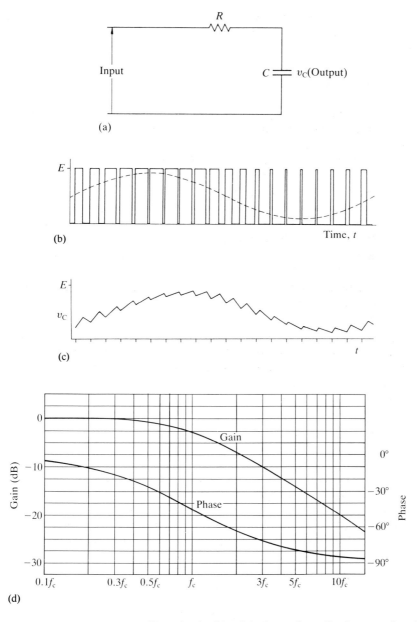

Fig. 5.20. (a) First-order CR filter circuit; (b) original waveform (broken curve) and its pulse-width modulated wave (input); (c) output waveform; (d) Bode diagram.

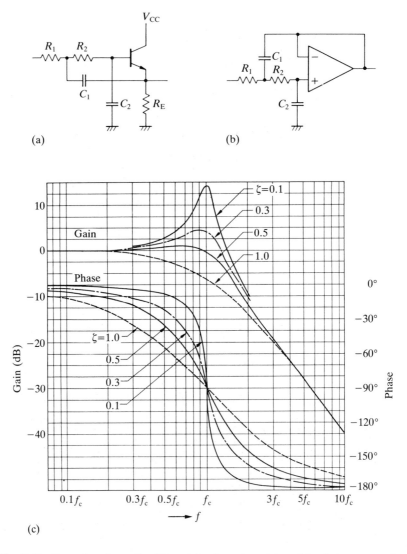

(a)   (b)

(c)

Fig. 5.21. Second-order active filter: (a) using a transistor; (b) using an operational amplifier; (c) Bode diagram.

recovered without much delay in phase. If the carrier frequency is 5 kHz, it will be attenuated by the factor of -52 dB, which means that in the demodulated waveforms the carrier frequency component is almost eliminated. Thus, a second-order active filter can be used as a suitable demodulator for observing waveforms.

(a)

(b)

Fig. 5.22. Photograph of an experimental d.c. servomotor system using a servo-amplifier usable in both linear and PWM modes.

5.7 **Examples of servo-amplifiers**

In this section, two servo-amplifiers will be presented. One is a transistor
amplifier that was designed by the author and his laboratory colleagues for
student experiments with microprocessor control of a d.c. motor. This is
unique in that the circuit can be used in either the linear or the PWM mode in
the voltage-control scheme. The second is a MOSFET PWM servo-
amplifier that was designed by two engineers who studied at the author's
laboratory, to be used in the current-control mode in practical applications.

5.7.1 *Experimental voltage-controlled servo-amplifier*

Figure 5.22 is a photograph of an educational-purpose position servo system
designed by the author and his colleagues. This is a unique voltage-controlled
scheme that can be operated either in the linear or PWM mode at a frequency
of about 20 kHz.

As shown in the block diagram of Fig. 5.23, the speed instructions can be
given from (1) a microprocessor via a digital-to-analogue converter, (2) an
external voltage signal, or (3) a manually adjustable voltage using a rheostat.

A simple choice in terms of principle for the speed sensor may be a d.c.
tachogenerator, as will be dealt with in the explanation of position control in
Chapter 10. However, a pulse generator is often used in low-cost equipment,
in spite of the complexity of the circuit. In this system, too, a pulse generator
is coupled to the motor shaft, and its two-channel signals are used to detect
the rotational speed and its direction.

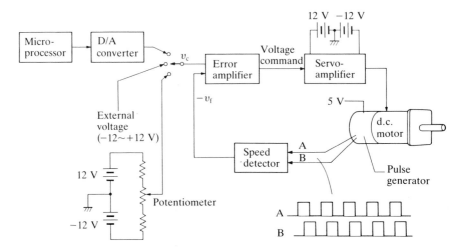

Fig. 5.23. Block diagram of the system shown in Fig. 5.22.

The block diagram of the speed/direction detector is given in Fig. 5.24 along with its functional diagram. The detailed circuit diagram is seen in Fig. 5.25, and part of the error amplifier and the power amplifier is shown in Fig. 5.26. Three operational amplifiers are used here for different purposes. The first amplifier, which receives the speed command v_c, forms a voltage-follower scheme to work as a buffer for the input signal. Since the input impedance of the operational amplifier is effectively infinite, the impedance in the D/A converter or the rheostat in the preceding stage does not affect the gain and phase characteristics in this stage and the speed command voltage v_c appears at point A also.

The speed feedback voltage v_f is applied to R_4 with opposite polarity to that of v_c. If R_4 and R_5 have the same resistance, the potential at point B is $(v_c - v_f)/2$ and this is applied to operational amplifier 2. Resistor R_2 and rheostat R_3 are used for fine offset adjustment to set the output voltage at

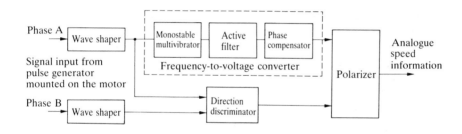

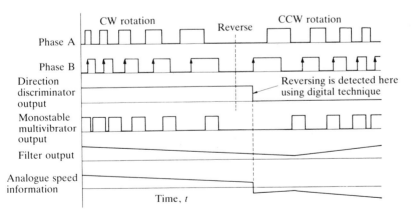

Fig. 5.24. Block diagram and function of the speed detector using a two-phase pulse generator.

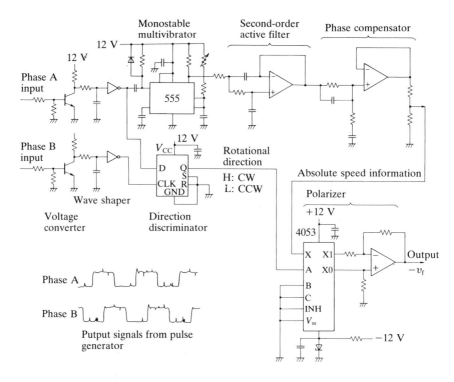

Fig. 5.25. Speed-detector circuit diagram. Two-phase pulse trains may be distorted as shown in this figure, but the noise components are eliminated in the wave-shaper stage before application to the monostable multivibrator using a timer chip 555.

operational amplifier 2 to zero when the command and feedback voltages are the same but of opposite polarity. Operational amplifier 2 is used as a voltage amplifier with first-order delay characteristics. The gain for the low-frequency component is R_7/R_6, which is 100 times or 40 dB in this example.

When switches S_1 and S_2 are turned to the PWM side, operational amplifier 3 works as a comparator, as explained in Figs. 5.9 and 5.17. It is obvious that the gain is -1. When the switches are turned to the LINEAR side, the amplifier works as a reversal amplifier whose gain is R_9/R_8, provided that the effects by other circuit parameters are negligible. Hence, the same resistance should be selected for R_8 and R_9 to make the gain also to be -1. R_{10}, which is 100 Ω here, is an auxiliary resistor to secure a stable operation of this stage.

The fundamental scheme of the power stage is that shown in Figs. 5.2 and 5.9. The main devices for driving the motor are Tr_3 and Tr_4. The other elements in the power stage are for over-current protection, and details are given in Section 5.7.3.

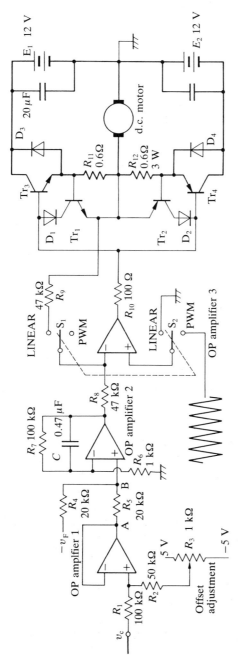

Fig. 5.26. Stages of the speed-error amplifier and power amplifier, which can be used either in the linear or PWM mode.

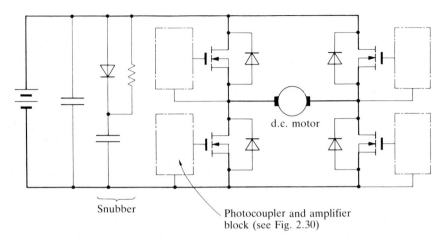

Snubber

Photocoupler and amplifier
block (see Fig. 2.30)

Fig. 5.27. An H-bridge using four MOSFETs. When three phases with a total of six MOSFETs are provided, it can be an inverter for an a.c. motor or a brushless d.c. motor drive.

Refer to Section 10.5.2 as for the details of the D–A converter used in Fig. 5.23 for converting the digital speed command generated by the microprocessor to an analogue command.

5.7.2 *MOSFET type*

An example of an H-bridge PWM amplifier using MOSFETs is given in Fig. 5.27. As with the bipolar transistor circuit of the MECHATRO LAB discussed in Chapter 1, photocouplers are used for interfacing between the TTL signal and the MOSFET gates, using the circuit configurations shown in Fig. 2.28 or 2.29 in Chapter 2.

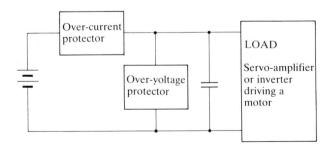

Fig. 5.28. Over-current and over-voltage protectors.

5.7.3 *Protection circuits*

In many bridge-type converters, servo-amplifiers, and also inverters to be discussed in Chapter 7, the overall current is often limited to a preset level. Furthermore, over-voltage protection is also needed in many applications. When the motor speed is reduced from high to low, the machine works as a generator, and current will be fed back to the capacitor placed between the power supply terminals. Owing to the charge stored in this capacitor, the voltage can rise enough to damage the transistors. Consequently, many converters have over-current and an over-voltage protector circuits in the arrangement shown in Fig. 5.28. Detailed examples of these are explained in Figs. 5.29 and 5.30.

The over-current protector used in the servo-amplifier shown in Fig. 5.26 is very useful, too. Let us consider what happens when the output voltage at operational amplifier 3 is positive and Tr_3 is carrying a motor current

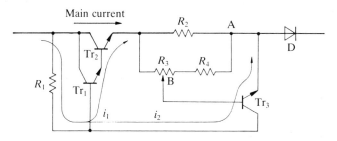

Main current

R_1	Resistor for restricting the base current of Tr_1
R_2	Resistor for detecting the main current
R_3 and R_4	Resistor for adjusting the maximum current
D	Diode needed to block feedback current

Fig. 5.29. Example of a current limiter. Normally i_1 flows through R_1 and the emitters of Tr_1 and Tr_2, which are in the Darlington connection, to turn these transistors ON. R_2 is of low resistance for sensing current, R_3 and R_4 are for adjustment of the voltage detected at point B. As long as the potential at B with respect to A, which is proportional to the main current, is less than 0.6 V, the normal condition is maintained. When the potential at B exceeds 0.6 V, the minimum forward voltage with respect to point A, then Tr_3 closes to cause current i_2 to flow. After this, i_1 tends to decrease because the impedance for i_1 is higher than i_2, and the Darlington transistors Tr_1 and Tr_2 become open, which preventing the main current from being higher than the set value that is given by

$$\text{Set current} = \frac{(0.6/R_2)(R_3 + R_4)}{\eta\, R_3 + R_4},$$

where η is the division factor of R_3. Provided that $R_1 = 10$ kΩ, $R_2 = 0.1$ Ω (5 W), $R_3 = 200$ Ω, $R_4 = 200$ Ω, $\eta = 0$ to 1, then the limit current is adjustable from 6 to 12 A.

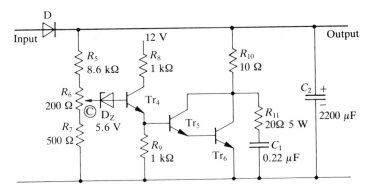

Fig. 5.30. Example of an over-voltage protector. When the load machine becomes a generator, a d.c. current is fed back from the power stage towards the converter. However, since diode D blocks this current it will flow into capacitor C_2. If this current is an instantaneous phenomenon, the rise in the capacitor voltage is low. However, if this current continues to flow, the potential across the capacitor C_2 eventually exceeds the transistor's voltage ratings. To prevent this, the potential at point C is compared with the breakdown potential of the Zener diode D_Z. When this potential exceeds the breakdown potential plus the base-to-emitter forward voltage, transistors Tr_4, Tr_5, and Tr_6 will all turn on to discharge C_2. When the potential at point C decreases and the Zener diode is brought back to the normal state, Tr_5 and Tr_6 are again operated. Thus, the potential applied to the power circuit is kept lower than a certain value given by

$$V_{max} = \frac{(V_Z + 0.6)(R_5 + R_6 + R_7)}{\xi R_6 + R_7}$$

where ξ is the division factor of R_6, which varies from 0 to 1. When the circuit parameters are provided as given in the figure, assuming that $V_Z = 5.6$ V, then V_{max} will be

$$V_{max} = 6.2 \times 9300/(500 \text{ to } 700) = 82 \text{ to } 115 \text{ V}.$$

The combination of R_{11} and C_1 comprises a snubber, needed to absorb a surge voltage that might occur when Tr_6 turns off.

supplied from E_1. Diode D_1 is for compensation of the forward voltage between the base and emitter in Tr_3. Resistor R_{11}, which is 0.6 Ω in this example, is placed for sensing the emitter current. Transistor Tr_1 is off as long as the potential drop in this resistor is below 0.6 V or the current is below 1 A. However, if the current exceeds 1 A, Tr_1 begins to turn on. As a result, the current that has been flowing to the base of Tr_3 will now bypass through D_1 and Tr_1, which implies that Tr_3 is brought into the linear region from the saturation region to restrict the collector current to 1 A. When the output from operational amplifier 3 is negative, D_2, Tr_2 and R_{12} work as the over-current protector for Tr_4.

Reference

1. Kenjo, T. and Nagamori, S. (1985). *Permanent-magnet and brushless DC motors*. Oxford University Press, Oxford.

6 Stepping-motor drive technology

The stepping motor, which is inherently adaptable to digital control, has recently been used in great numbers in office automation equipment and microcomputer peripherals. As details of stepping motors are dealt with in one of the author's earlier books,[1] fundamental types of power circuits for stepping motors will first be discussed here, in conjunction with the circuits for other types of motors. The use of microprocessors will then be discussed quoting some examples; such as the use of a Z80 for acceleration/deceleration control and use of a single-chip microprocessor for multiple motors. Finally, the drive technology of a stepping motor similar to the drive of a brushless d.c. motor will be presented.

6.1 Three basic types of stepping motor

There are three distinctive types of stepping motor in current use. The details are described in Chapter 2 of Reference [1], but the topic will be surveyed and some modern motors will be described here.

(1) *Variable-reluctance motors.* The elementary stepping motor is the variable-reluctance (VR) motor. The basic construction of the single-stack VR motor and its principle were explained in Chapter 1. Multistack variable-reluctance motors were once used as the prime movers in numerical control machines, but they have recently been replaced with brush or brushless d.c. servomotors. Single-stack VR motors are still used in some office machines.

(2) *Hybrid motors.* A recent trend is miniaturization of motors or actuators used in office automation equipment. Use of powerful rare-earth magnets have made the machines very compact. Hybrid stepping motors using disc magnets are extensively employed in high-accuracy positioning applications. The photographs of Fig. 6.1 show three examples of miniature hybrid motors used for magnetic-head positioning in floppy/hard disc drives.

Type (a) possesses four poles* and each pole has four teeth; this motor completes one revolution in 100 steps. Thus the step angle is 3.6°. Type (b),

---

*This use of the term 'pole' in stepping motors differs from the concept of magnetic poles. A pole is a large tooth that has smaller teeth, and around which coils are wound.

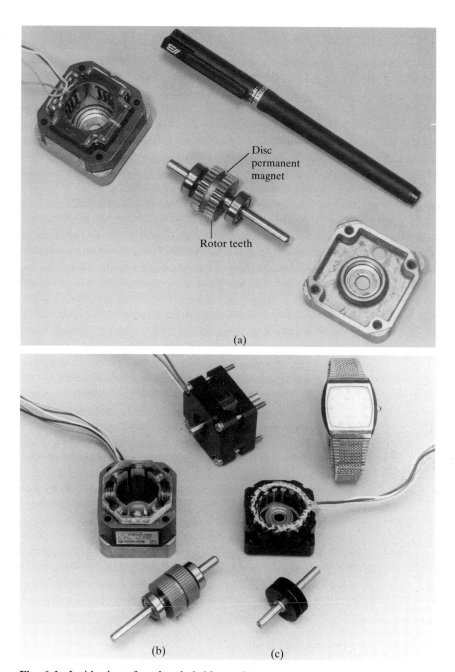

Fig. 6.1. Inside view of modern hybrid stepping motors to be used for magnetic head positioning in floppy/hard disc drives. Motor (a) is a four-poled bifilar-wound motor having 25 teeth on its rotor to yield a 3.6° step; motor (b) is a bipolar-operated, eight-poled two-phase type; motor (c) is 16-poled bipolar-operated motor having 100 teeth on its rotor, to yield a 0.9° step.

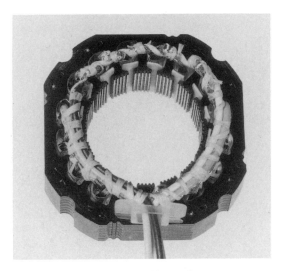

Fig. 6.2. Fine teeth on stator poles in a hybrid stepping motor whose step angle is 0.9°
(shown in Fig. 6.1(c)).

whose step angle is 1.8°, possesses eight poles having four teeth. In type (c)
motors there are sixteen poles on its stator and 100 teeth on the rotor. Figure
6.2 shows the shapes of the poles and teeth of this motor, whose step angle is
0.9°.

A unique type is the five-phase hybrid motor. As illustrated in Fig. 6.3, this
machine has ten stator poles, and the typical number of teeth on the rotor is
72, which produces a step angle of 0.72°.

Figure 6.4 shows a linear motor designed on the hybrid principle to be controlled in a brushless d.c. scheme in an image scanner. Linear stepping
motors are also employed in manufacturing machines in various fields.

(3) *Claw-pole permanent-magnet motors.*   Another type of stepping motor
using a permanent magnet is the claw-pole permanent-magnet motor shown
in Fig. 6.5. The claw-pole motor is a low-cost type, while the hybrid motor
features high positioning accuracy and quick responses.

6.2 **Monofilar, bipolar, and bifilar**

The basic drive circuit shown in Fig. 1.8 in Chapter 1 is the simplest, excepting those for single-phase motors used in wrist-watches. As each pole produces only a North or South magnetic pole, this drive is referred to as the
'unipolar' or 'monopolar drive', and is employed for VR motors.

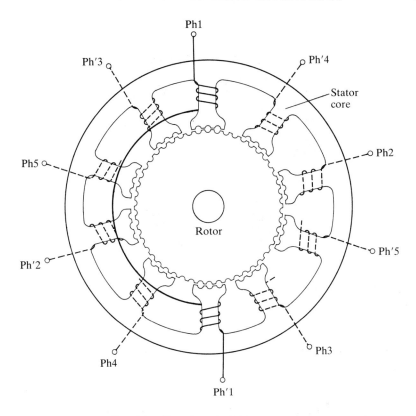

Fig. 6.3. Five-phase hybrid stepping motor.

Fig. 6.4. Linear hybrid stepping motor used as a brushless servomotor. (By courtesy of Matsushita Electric Industrial Co., Ltd.)

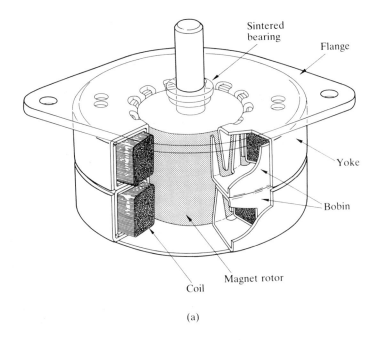

(a)

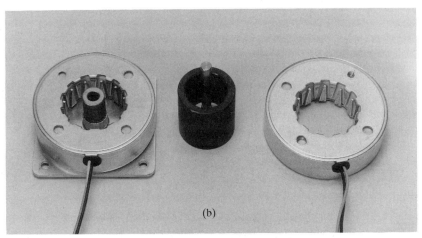

(b)

Fig. 6.5. Claw-poled permanent-magnet stepping motor; (a) cutaway view; (b) disassembled motor.

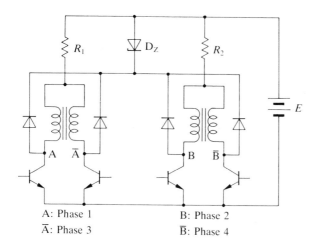

A: Phase 1
$\overline{A}$: Phase 3

B: Phase 2
$\overline{B}$: Phase 4

Fig. 6.6. An example of drive circuit of the bifilar-wound four-phase motor.

There are basically two arrangements for a pole to create the North and South poles in turn, and these methods are extensively employed in the operation of hybrid and claw-pole motors.

(1) *Bifilar-wound motor.* Though not shown clearly in the picture of stator (a) shown in Fig. 6.1, two wires having different colours coated in enamel are wound together on a pole. The winding arrangement is as shown in Fig. 6.6, the coils of Phase 1 (or A) and Phase 3 (or $\overline{A}$) are mounted on the same two poles I and III, and the coils of Phases 2 and 4 (or B and $\overline{B}$) are on Poles II and IV.

When these wire terminals are connected to a circuit as shown in Fig. 6.6, Phase 1 and Phase 3 are used to excite opposite magnetic poles. Similarly, if Phase 2 yields the North pole, Phase 4 produces the South pole. It should be noted here that this motor can be regarded as a two-phase motor when the combination of Phase 1 and Phase 3 can be regarded as Phase A, and Phases 2 and 4 form Phase B.

(2) *Bipolar drive.* Another choice is the H-bridge scheme shown in Fig. 6.7(a); an example of the coil arrangement is illustrated in Fig. 6.7(b) for an eight-poled motor. Phase A consists of odd-numbered poles, and Phase B consists of even-numbered poles. The differences in one-phase-on, two-phase-on, and half-step one/two-phase-on operations for the bipolar-scheme drive are illustrated in Fig. 6.8. for the four-phase motor.

It is known that the five-phase hybrid stepping motor provides excellent dynamic performance when this is operated with five H-bridge circuits using

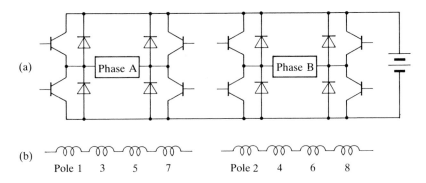

Fig. 6.7. (a) A set of H-bridge circuits for the bipolar operation of a hybrid or a claw-poled PM stepping motor; (b) coil arrangement for an eight-poled motor.

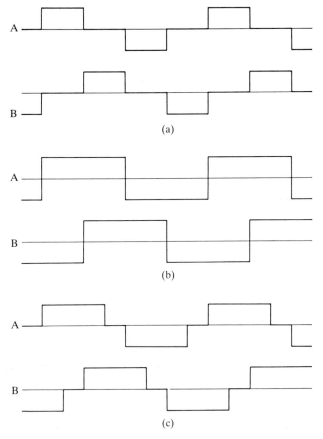

Fig. 6.8. Differences between (a) one-phase-on, (b) two-phase-on, and (c) one/two-phase-on operation in the bipolar drive.

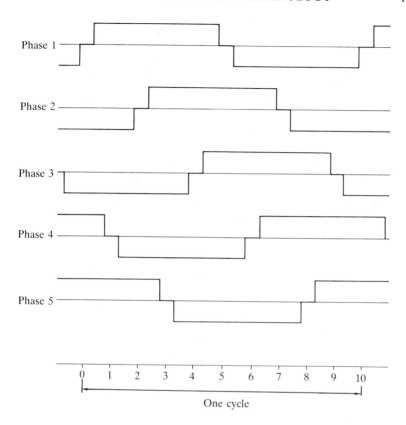

Fig. 6.9. Ideal current variation in the 4/5-phase-on operation of a five-phase motor.

20 power transistors or MOSFETs. The ideal current waveforms in 4/5-phase-on operation are shown in Fig. 6.9.

6.3 High-speed operation

In the drive of a stepping motor, the time lag of the current behind voltage is often a serious problem and some corrective measures must be taken. Details have been given in Chapter 2 of Reference [1], and some modern methods are briefly reviewed here.

(1) *Use of resistors and Zener diodes.* As seen in the circuit of Fig. 6.14, which will be used in the next section, a resistor is connected in series with the parallel connection of two phases to reduce the electrical time constant, and a

Zener diode is used in series with the flyback diode to suppress the transient current rapidly after turning off.

(2) *Dual-voltage drive.*   As seen in Fig. 6.10, two power supplies, one high voltage and the other low voltage, are provided: the higher-voltage supply is used only for the beginning period of the ON interval until the current reaches the rated value, and then Tr_1 or Tr_2 is turned off so that the motor current is supplied from the lower-voltage power supply.

(3) *Chopper drive.*   The principle of the chopper or PWM drive is illustrated in Fig. 6.11, where the load is simulated by a combination of a resistance and a reactance.

Transistor Tr_1 is repeatedly turned on and off at a high frequency to adjust automatically the current at a level determined as the reference voltage applied to the negative terminal of the comparator. A d.c. level superimposed using a triangular waveform is applied to the reference terminal. When the current pickup voltage v_o is higher than the reference level, the T_{on} period in Tr_2 decreases; on the other hand, when the pickup voltage is lower than the reference level, T_{on} increases to adjust the current at the level. In this scheme, the PWM frequency is the same as the triangular wave frequency superimposed on the d.c. reference voltage.

Figure 6.12 shows an application of this principle to driving a three-phase VR stepping motor.

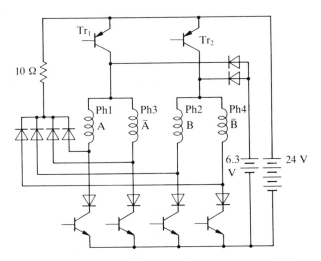

Fig. 6.10. Dual-voltage bifilar drive of a four-phase hybrid motor.

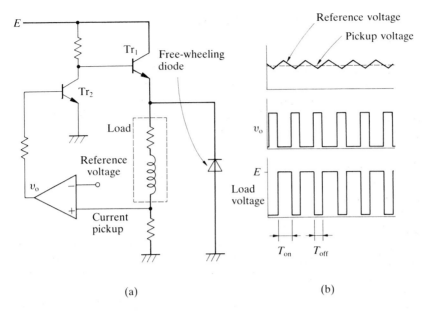

(a) (b)

Fig. 6.11. Principle of the chopper or PWM drive of a stepping motor: (a) fundamental circuit; (b) waveforms.

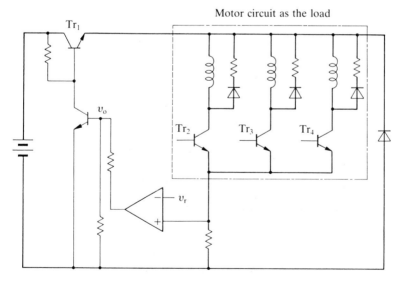

Fig. 6.12. PWM drive circuit for a three-phase VR motor.

6.4 Application of microprocessors

Microprocessors have been used in large numbers and in various ways in controlling stepping motors. Although a single microprocessor can control several stepping motors at the same time, it is not always practical to have it do so, because the program flexibility is decreased greatly. Basically, the normal procedure is to use one microprocessor for one motor, and when many motors are to be controlled in synchronism or in a certain sequence, their microprocessors are supervised by a host microprocessor. Some typical examples follow.

6.4.1 Acceleration, slewing, and deceleration using one microprocessor

In Chapter 7 of Reference [1], an example is given of using an 8080A microprocessor for the acceleration/deceleration drive. In this example, however, the maximum number of steps in a movement from start to finish is limited to 256. It is clearly possible to improve the software to deal with a motion up to 32767 steps in either direction by using two bytes (16 bits) for storing one

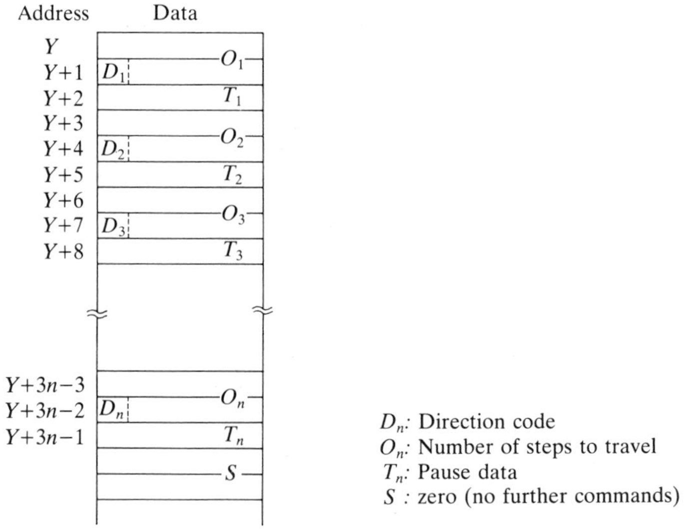

D_n: Direction code
O_n: Number of steps to travel
T_n: Pause data
S : zero (no further commands)

Fig. 6.13. Image diagram of motion-command data. Three bytes are used for one motion. The number of steps is given by the 15-bit data; the first byte (8 bits) of each section is the lower-digit data, and the lower seven bits in the second byte are the higher digits of the step data. The MSB (most significant bit) in the second byte is the direction code. The third byte is related to the pause before the next motion is commanded.

motion command, as illustrated in Fig. 6.13. Here three bytes of memory are used for instructing one motion. The first two are for the number of steps, and the last is the datum for a pause before the next motion is instructed. The MSB (most significant bit) in the second byte is for the direction instruction: 0 for CW and 1 for CCW.

However, since the kinds of machine language instruction are limited and there are fewer registers with the 8080A or 8085, such programs can be very complicated. The use of a Z80, which is regarded as an advanced version of the 8080A, is reasonable for this purpose, because the algorithm developed in Reference [1] can easily be improved by using some additional instructions. The details have been described in one of the author's Japanese books.[2] The outline is stated here.

Figure 6.14 shows an example of connection of the Z80 processor to a drive circuit to which a bifilar-wound motor is connected. The speed profiles employed in this system are illustrated in Fig. 6.15. For the sake of simplicity, acceleration and deceleration are carried out symmetrically. This means that same step intervals are prepared, and the data are utilized forwards in the acceleration and backwards in the deceleration. When the number of steps

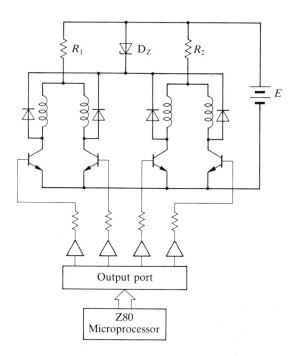

Fig. 6.14. Connecting a Z80 microprocessor and a drive circuit.

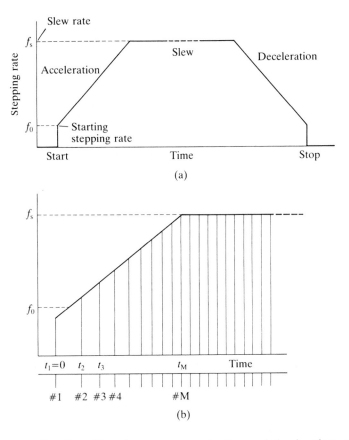

Fig. 6.15. (a) Speed profile of acceleration, slewing, and deceleration and step intervals. (b) Step timings in acceleration are determined so that each rectangular area equals to one step.

exceeds a certain value, a slew region will appear between acceleration and deceleration.

In this example, the motor is accelerated from a starting stepping rate of 200 Hz (steps per second) and attains a slew rate of 2000 Hz at the 200th step.

The registers are used for the following purposes:

A and A′   Various eight-bit logical and arithmetic operations and storing
           the rotational direction code
B C   Counting the completed steps
D E   Storing the remaining steps
H L   Storing step interval data, and arithmetic computation of
           16-bit data

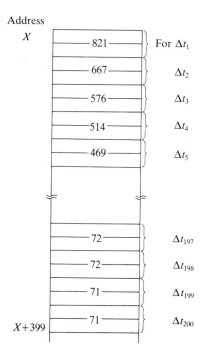

Fig. 6.16. Step interval data Q_n stored in a memory area. Two bytes are used for each step interval.

IX   Specifying the address storing step interval data
IY   Specifying the address storing the drive command

The general flowchart of this drive is as shown in Fig. 6.17, and its assembly language program listing is given in Table 6.1.

In this example the step interval Δt_n is related to the step interval data Q_n by the following equation:

$$\Delta t_n = \frac{24Q_n + 289}{f_c} , \qquad (6.1)$$

where f_c is the clock frequency for operation of the CPU, being 4 MHz in this case. The factors '24' and '289' are related to the number of processing states of each command used in this program as follows:

Step command generation routine	59 states
Counter routine	30 states
Address specification routine	148 states
Step interval adjusting routine	$24Q_n + 52$ states
Total	$24Q_n + 289$ states.

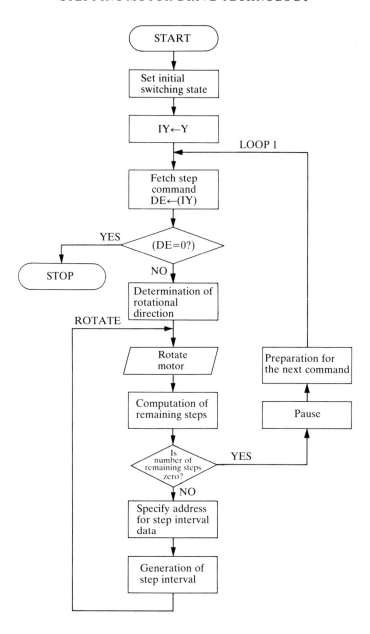

Fig. 6.17. General flowchart of the acceleration/deceleration drive.

Table 6.1. Program listing for acceleration/deceleration drive of a four-phase stepping motor.

```
         ORG     9000H

N        EQU     200        ;Number of steps before reaching slew
R        EQU     399        ;2*N-1

START:   LD      B,33H      ;(7)   Load B with 00110011B
         LD      C,0FDH     ;(7)   Load C with output port address FCH
         OUT     (C),B      ;(12)  Output data in B via port FCH
         EXX                ;(4)   Keep data in BC by exchanging with BC'
         LD      IY,Y       ;(14)  Load IY with the initial address for step
                            ;        command data

LOOP1:   LD      E,(IY+0)   ;(19)  Load DE with number of steps to travel
         LD      D,(IY+1)   ;(19)
         LD      A,D        ;(4)
         OR      E          ;(4)   If DE=0, then go to STOP
         JP      Z,STOP     ;(10)
         LD      A,D        ;(4)
         ADD     A,0        ;(7)   If number of steps is minus,
         JP      M,MINUS    ;(10)  then go to MINUS

PLUS:    NEG                ;(8)   Adjust time
         NEG                ;(8)   Adjust time
         LD      A,0        ;(7)   Since positive, Load A with 0(zero)
         JP      CLEAR      ;(10)  and go to CLEAR

MINUS:   CPL                ;(4)   Now number of steps is minus. Take its
         LD      D,A        ;(4)   absolute value and
         LD      A,E        ;(4)   load it in DE
         CPL                ;(4)
         LD      E,A        ;(4)
         INC     DE         ;(6)
         LD      A,-1       ;(7)   Load A with -1

CLEAR:   LD      BC,0       ;(10)  Clear BC

ROTATE:  EXX                ;(4)   Retrieve exciting state data and port address
         ADD     A,0        ;(7)   Make flags affected
         JP      M,RIGHT    ;(10)  If MSB is 1, then go to RIGHT, else LEFT
LEFT:    RLC     B          ;(8)   Rotate data in B left
         JP      STEP       ;(10)  and go to STEP

RIGHT:   RRC     B          ;(8)   Rotate data in B right
         JP      STEP       ;(10)  and go to STEP

STEP:    OUT     (C),B      ;(12)  Output data in B to step the motor
         EXX                ;(4)   Store these data in BC'
         EX      AF,AF'     ;(4)   Keep data by exchanging AF and AF'
```

Table 6.1. *continued*

```
COUNT:   INC   BC          ;(6)   Count up a step
         DEC   DE          ;(6)   Count down a step
         LD    A,D         ;(4)
         OR    E           ;(4)   If the remaining step (DE) is zero
         JP    Z,LOOP3     ;(10)  then go to LOOP3

         LD    IX,X-1      ;(14)  Load IX with X-1, where X is the starting address
                          ;       for the acceleration data
         LD    H,D         ;(4)
         LD    L,E         ;(4)
         SCF               ;(4)   Set carry flag
         CCF               ;(4)   and invert it (now CY=0)
         SBC   HL,BC       ;(15)  HL-BC (or compare DE and BC)
         JP    M,NSUBDE    ;(10)  If minus, go to NSUBDE

NSUBBC:  LD    HL,N        ;(10)  Load HL with data N being the steps before
                          ;       reaching slew and now 200
         SCF               ;(4)   Set carry flag
         CCF               ;(4)   and invert it
         SBC   HL,BC       ;(15)  If N<BC
         JP    M,SLEW      ;(10)  then go to SLEW

ACCEL:   ADD   IX,BC       ;(15)
         ADD   IX,BC       ;(15)  Add 2*BC-1 to IX
         DEC   IX          ;(10)
         JP    WAIT        ;(10)  and go to WAIT

NSUBDE:  LD    HL,N        ;(10)  Load HL with N
         SCF               ;(4)   Set carry flag
         CCF               ;(4)   and invert it
         SBC   HL,DE       ;(15)  If N>BC
         JP    P,DECEL     ;(10)  then go to DECEL
SLEW:    EXX               ;(4)   Keep data by exchanging DE and DE'
         LD    D,0         ;(7)   Adjust time by using this command
         LD    DE,R        ;(10)  Load DE with data M being 399
         ADD   IX,DE       ;(15)  Add DE to IX
         EXX               ;(4)   Retrieve DE
         JP    WAIT        ;(10)  Go to WAIT

DECEL:   ADD   IX,DE       ;(15)
         ADD   IX,DE       ;(15)  Add 2*DE-1
         DEC   IX          ;(10)
         JP    WAIT        ;(10)  and go to WAIT

WAIT:    LD    L,(IX+0)    ;(19)  Load HL with pulse interval data
         LD    H,(IX+1)    ;(19)
LOOP2:   DEC   HL          ;(6)
         LD    A,H         ;(4)
         OR    L           ;(4)   Generate a time based on this data
         JP    NZ,LOOP2    ;(10)
         EX    AF,AF'      ;(4)   Retrieve AF
         JP    ROTATE      ;(10)  Go to ROTATE
LOOP3:   INC   IY          ;(10)  Specify time data address
         INC   IY          ;(10)  by adding 2 to IY
         LD    A,(IY+0)    ;(19)  Load register A with the data in (IY+0)
```

Table 6.1. *contd.*

```
LOOP4:   LD      D,195      ;(7)
LOOP5:   LD      E,145      ;(7)
LOOP6:   DEC     E          ;(4)
         JP      NZ,LOOP6   ;(10) Spend a time as a function of
         DEC     D          ;(4)  data in A
         JP      NZ,LOOP5   ;(10) (About A times 100ms)
         DEC     A          ;(4)
         JP      NZ,LOOP4   ;(10)

         INC     IY         ;(10) Increment IY
         JP      LOOP1      ;(10) Go to LOOP1

STOP:    HALT               ;     Stop CPU

;***   Acceleration/deceleration data table   ***

X:       DW      821,667,576,514,469,433,404,381,361,343
         DW      328,315,303,292,283,274,266,258,252,245
         DW      239,234,229,224,219,215,211,207,203,200
         DW      196,193,190,187,185,182,179,177,174,172
         DW      170,168,166,164,162,160,158,156,155,153
         DW      152,150,148,147,146,144,143,142,140,139
         DW      138,137,135,134,133,132,131,130,129,128
         DW      127,126,125,124,123,122,122,121,120,119
         DW      118,117,117,116,115,114,114,113,112,112
         DW      111,110,110,109,108,108,107,107,106,105
         DW      105,104,104,103,103,102,101,101,100,100
         DW      99,99,98,98,98,97,97,96,96,95
         DW      95,94,94,93,93,93,92,92,91,91
         DW      91,90,90,90,89,89,88,88,88,87
         DW      87,87,86,86,86,85,85,85,84,84
         DW      84,83,83,83,82,82,82,82,81,81
         DW      81,80,80,80,80,79,79,79,78,78
         DW      78,78,77,77,77,77,76,76,76,76
         DW      75,75,75,75,74,74,74,74,74,73
         DW      73,73,73,72,72,72,72,72,71,71

;***   Drive commands   ***

         ORG     0A000H

Y:       DW      8000       ;CW 8000 steps
         DB      15         ;1.5 sec pause
         DW      -8000      ;CCW 8000 steps
         DB      17         ;1.7 sec pause
         DW      10000      ;CW 10000 steps
         DB      16         ;1.6 sec pause
         DW      -10000     ;CCW 10000 steps
         DW      3          ;3 sec pause
         DW      0          ;No more commands

         END
```

Theoretically, the step intervals Δt_n for $n = 1$ to 199 are give by the equation

$$\Delta t_n = \frac{2}{(g^2 + 2nb)^{\frac{1}{2}} + [g^2 + 2(n-1)b]^{\frac{1}{2}}} \tag{6.2}$$

where

$$g = f_0 - b/2f_0 \tag{6.3}$$

f_0 = starting stepping rate

b = acceleration (steps/s²) given by

$$b = \frac{2(f_s^2 - f_0^2)}{[(2M - 3)^2 + (f_s/f_0)^2 - 1]^{\frac{1}{2}} + (2M - 3)} \tag{6.4}$$

with

$M = 200$

f_s = slew rate.

Hence, the Q_n are given by the equation

$$Q_n = (4 \times 10^6 \times \Delta t_n - 289)/24. \tag{6.5}$$

In practice, when storing these data in the memory area starting with label X, their integer approximated quantities must be used.

6.4.2 Control of several stepping motors

When a number of stepping motors are to be driven in a system, one microprocessor is used for each motor and these processors are controlled by one host processor. Two given examples follow.

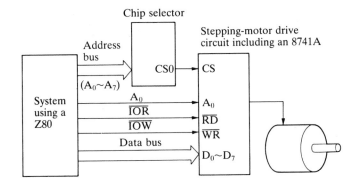

Fig. 6.18. Using a single-chip microprocessor supervised by a host processor.

(1) *Using one Z80 and one or more 8741As.* First we shall briefly look at the case, discussed in a Japanese book[3] by the author and a colleague, in which one or more single-chip microprocessors 8741A are supervised by a host Z80 as shown in Fig. 6.18. The 8741A is a single-chip processor suitable as a universal peripheral interface device in use with a higher-performance processor. The roles of the two processors are then as follows:

Host CPU Z80 — Read the motion-command data stored in the memory area and transmit to the 8741A

Peripheral CPU 8741A — Interpret the data transmitted from the host CPU, and transmit excitation data to the power circuit using an eight-bit rotation command at an arranged set of timings.

The detailed connections between the 8741A and the drive circuit are shown in Fig. 6.19. When several stepping motors are controlled in a multi-axis configuration, the system shown in Fig. 6.20 can be used. In this system, however, the maximum number of motors is eight.

(2) *Camera stand for special effects.* An interesting application of the multi-axis drive of stepping motors is a camera system for special effects in animation. In the example shown in Fig. 6.21, which was produced by Animation Staff Room, Tokyo, 16 stepping motors of five-phase hybrid type

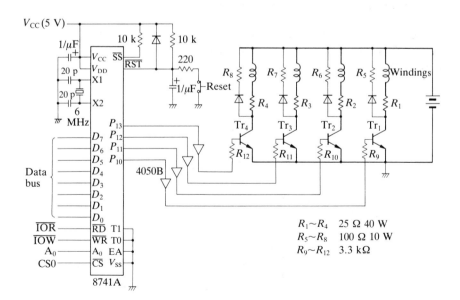

Fig. 6.19. Interfacing between 8041A and motor drive circuit.

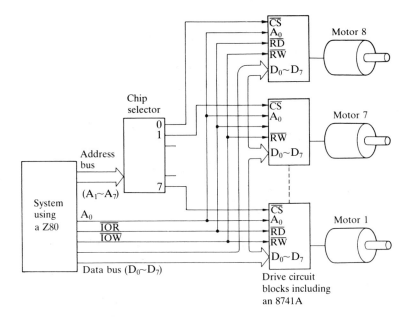

Fig. 6.20. System diagram of multi-axis control of stepping motors.

were used.[4] For each stepping motor, a Z80A processor is used, and 16 Z80As, are supervised by another Z80A to accelerate and decelerate the motors synchronization to one another.

In an application in which five-phase motors are controlled, it is not always reasonable to use the CPU itself to rotate a group of five bits, because a combination of several instructions is necessary and this results in long execution time. Usually, a dedicated shift-register chip is used, which receives a command from the CPU and generates a switching code to step the motor.

6.5 Brushless direct-current drive of stepping motors

A closed-loop adaptive control system using a four-phase hybrid motor was discussed in Chapter 7 of Reference [1]. In this system the microcomputer, which is a system consisting of a microprocessor and memory, learns the optimum timing for initiating deceleration from a series of trial-and-error experiments. Although it was confirmed that this technique is effective when the load is almost constant and only a limited number of different distances to the target occur, it was also true that this approach was not always the best when the load conditions varied frequently.

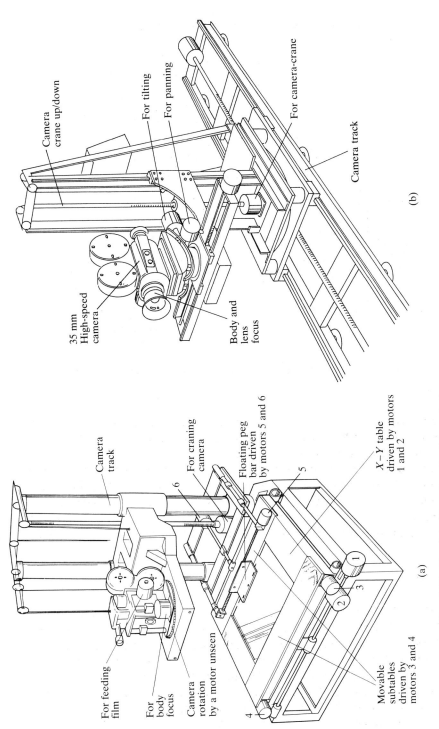

Fig. 6.21. Special effects stands: (a) vertical camera system; (b) horizontal camera system.

Camera track

6

For craning camera

Floating peg bar driven by motors 5 and 6

5

$X-Y$ table driven by motors 1 and 2

For feeding film

For body focus

Camera rotation by a motor unseen

Movable subtables driven by motors 3 and 4

1

3

2

4

(a)

Camera crane up/down

For tilting

For panning

For camera-crane

Camera track

35 mm High-speed camera

Body and lens focus

(b)

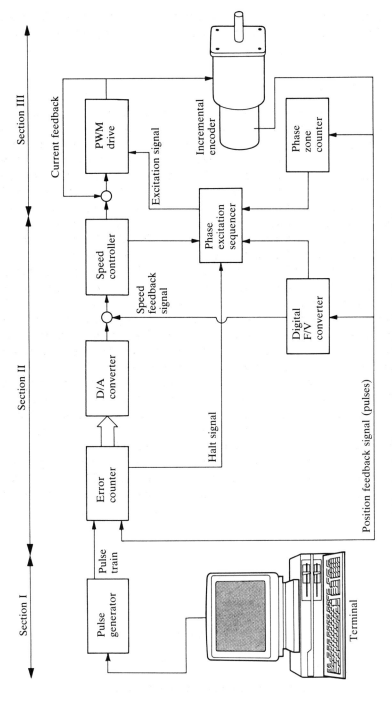

Fig. 6.22. System diagram for a brushless d.c. drive of a stepping motor.

Another approach[5] to closed-loop control of stepping motors is described here. This is a method compatible with a servo system using a conventional or brushless d.c. motor.

6.5.1 *System configuration*

Figure 6.22 shows the block diagram employed in this system. Section I of this system is the pulse generator or controller that generates pulse trains to govern the speed profile from start to the final target position. The same controller as in a d.c. servo system can be used here. Section II covers the control electronics. The basic configuration of this portion is similar to that part in a d.c. servomotor system, to be dealt with in Chapter 10, and generates speed commands as a function of position error or the difference between the number of pulses generated from the controller and the number of pulses from the position sensor. Section III is the portion of the closed-loop control to be dealt with here. The motor is a four-phase hybrid motor.

The functions of Sections I and II are unique in the following two respects. (1) In a d.c. motor, commutation is implemented in a fixed pattern by the mechanical relation between the brushes and commutator. In this system, however, commutation from one phase to another is governed by the phase excitation sequences according to position feedback signals, speed information, direction commands, and other items of information to be described later. (2) When the target position is reached, the phase excitation sequencer generates signals to cause the PWM drive to produce a high holding torque to halt the rotor. This means that the restoring torque at the final position is produced by the tendency of groups of teeth on the stator and the rotor to pull back into alignment under excited phase(s). It should be noted that a complex feedback mechanism is not necessary for producing stiffness.

The test motor has bifilar winding arrangements, as shown in Fig. 6.23, and is driven as a four-phase motor in the circuit of Fig. 6.27 in the half-step excitation mode. That is, if one of the four phases is excited at a given moment, two phases will be excited next, and thus one-phase-on and two-

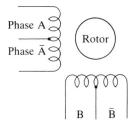

Fig. 6.23. Phase connections in the motor.

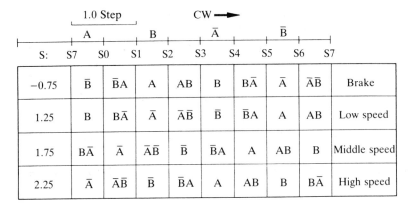

	S7	S0	S1	S2	S3	S4	S5	S6	
−0.75	B̄	B̄A	A	AB	B	BĀ	Ā	ĀB̄	Brake
1.25	B	BĀ	Ā	ĀB̄	B̄	B̄A	A	AB	Low speed
1.75	BĀ	Ā	ĀB̄	B̄	B̄A	A	AB	B	Middle speed
2.25	Ā	ĀB̄	B̄	B̄A	A	AB	B	BĀ	High speed

Fig. 6.24. Relation between lead angles, equilibrium positions, switching points and phases to be excited. Note: * indicates equilibrium positions; S indicates switching points or positions at which switching occurs.

phases-on alternate. The excitation sequence is determined by the rule given in Fig. 6.24, which gives the relation between lead angles, equilibrium positions, and phases to be excited. A definition of the lead angle is given in Chapter 7 of reference [1], but for this particular case it is explained as follows.

Let us suppose that the motor is now passing a region between S7 and S0 in a clockwise direction, following the excitation sequence in the third row in the table. According to the table, phase B and phase Ā are excited. When the rotor reaches switching point S0, phase B is de-energized, and so only phase Ā is now excited. The distance from S0 to the equilibrium position for this excitation state is 1.75 steps, which is the lead angle in this case. Thus, each phase is turned ON/OFF at a position earlier than the equilibrium position of the new excited state by this lead angle. In the second row, switching occurs at positions 1.25 steps earlier than equilibrium positions. At low speeds, the 1.25-step lead angle produces the highest torque, but the optimum lead angle increases with speed. Thus, sequential excitation with a lead angle of 1.25 steps or larger makes a four-phase stepping motor work in the 'motoring' mode. A lead angle of −0.75 step is suitable for deceleration or braking, because it develops an effective retarding torque.

6.5.2 Functions in detail

The functions of each block are as follows.

(1) *Position sensor.* The position sensor mounted on the rotor shaft is a two-channel optical encoder incorporating a home-position output, which

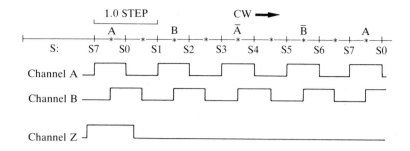

Fig. 6.25. Relation between encoder output signals, equilibrium positions, and switching points.

generates 200 pulses per channel per revolution. The relationship between the output waveforms, equilibrium positions, and switching points is illustrated in Fig. 6.25.

(2) *Error counter and D/A converter.* Position error, which is the difference between the commanded position and the actual position, is first counted in this stage, and the velocity command is then computed from the error. Finally, the digital command is converted to an analogue value. An eight-bit up/down counter and an 8748 CPU are used for counting error from -2^{23} to $+2^{23} - 1$ in units of ½ step.

(3) *Digital F/V converter.* A custom digital LSI is used for detecting the motor speed. The digital output is converted into an analogue voltage, which is used (i) as the speed feedback signal supplied to the speed controller, and (ii) as the information for changing the lead angle in the phase excitation sequencer.

(4) *Speed controller.* The analogue speed command is compared with the detected speed. The difference is amplified to be used as the current command for the next stage. The polarity of the speed error is also detected here and fed to the phase excitation sequencer.

(5) *Phase excitation sequencer.* At this stage the phase(s) to be excited are determined by the following eight-bit information.

1. *Rotor position* (3 bits). There are eight different regions in one excitation cycle. This information is generated in the phase zone counter.
2. *Speed range* (1 bit). In the experimental system, two different lead angles are used to develop accelerating torque: 1.75-step lead angle in the range below 1700 r.p.m., and 2.25-step lead angle in the higher ranges.

3. *Direction command* (1 bit). This is either CW or CCW.

4. *Master reset* (1 bit). When a master reset is executed, only phase A is excited.

5. *Speed error polarity* (1 bit). When (i) the commanded direction is CW and the speed error (= speed command minus actual speed) is positive, or (ii) the direction is CCW and the error is negative, the lead angle is set to 1.75 or 2.25 steps to drive the stepping motor in the motoring mode. On the other hand, (iii) if the commanded direction is CCW and the error is positive, or (iv) the direction is CW and the error is negative, the lead angle is set to −0.75 step to brake the machine.

6. *Halt* (1 bit). When the target position is reached, a signal is released to halt the rotor in the holding mode. The relation between the rotor position and phase(s) to be excited in this mode is given in Fig. 6.26.

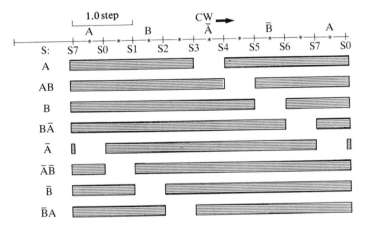

Fig. 6.26. Relation between rotor positions and phases to be excited in the holding mode. If the holding mode is set when phase A is excited, the rotor will stay with this excitation state in the shaded zone in the first row. However, if the rotor travels out of this region, the holding mode will be released. Likewise, the second row indicates the region in which phases A and B are excited in the holding mode. It should be noted here that at the edges of any shaded zone a strong restoring torque will be produced to pull the rotor back to the equilibrium position, which is the centre of each zone. (* indicates equilibrium positions; S indicates switching positions.)

A ROM chip is used to store the relationship between these items of information and the phases to be excited.

7. *PWM drive.* The circuit used is the unipolar scheme shown in Fig. 6.27. The MOSFETs are used for PWM operation at a frequency of 20 to 25 kHz, and four bipolar transistors are for sequential excitation. Bipolar drive is also usable.

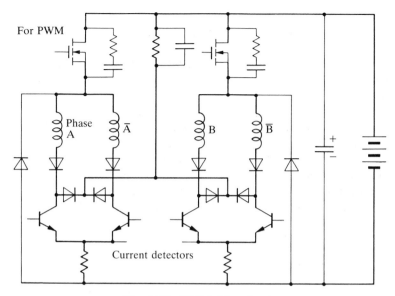

Fig. 6.27. P W M drive circuit.

6.5.3 *Speed pattern*

The acceleration characteristics are illustrated in Fig. 6.28 for two cases, in which the only load is the encoder disc. In the first case, when the lead angle is fixed at 1.75 steps and remains unchanged, the maximum speed is about 3800 r.p.m. In the second case, where the lead angle is switched to 2.25 steps when the speed exceeds 1700 r.p.m., the speed can build up to 6000 r.p.m., and the time needed to accelerate the motor up to 3000 r.p.m. is less than 300 ms.

However, if a 1.25-step lead angle is used for acceleration at low speeds in addition to the 1.75 and 2.25-step lead angles for higher speed ranges, the accelerating time will be much shorter and comparable with that of a system that uses a conventional or brushless d.c. motor. It should be noted that about 300 ms are needed to ramp a similar load up to 300 r.p.m. with typical sophisticated open-loop control.

If the speed command is given by a square-root function of position error, as shown in Fig. 6.29, the speed profile for one motion is like that shown in Fig. 6.30, similar to the d.c. servomotor case. For the mathematical explanation for these characteristics, the reader is referred to Section 10.3. However, in many practical applications of positioning controls, the controller, which is practically a pulse generator, is put prior to the drive system to provide suitable pulse timing for acceleration/deceleration in reference to other axes.

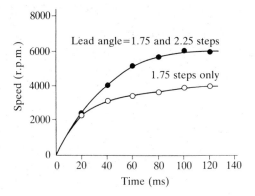

Fig. 6.28. Acceleration characteristics.

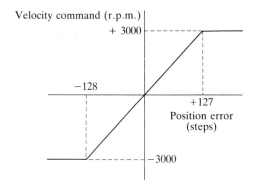

Fig. 6.29. Velocity commands versus position errors.

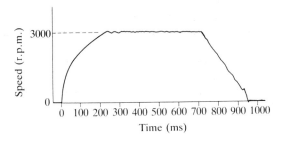

Fig. 6.30. Speed profile of a motion.

Fig. 6.31. Control electronics.

The photograph in Fig. 6.31 is of laboratory-made electronics for this drive system.

References

1. Kenjo, T. (1984). *Stepping motors and their microprocessor controls.* Oxford University Press, Oxford.
2. Kenjo, T. (1986). *Mechatronics controls using Z80/8085.* (In Japanese). Chapter 9. Sogo Electronics Publishing Company, Tokyo.
3. Kenjo, T. and Takahashi, H. (1986). *Microprocessor controlled mechatronics* (In Japanese), Chapter 3. Technical Research Center Co., Ltd., Tokyo.
4. Fujii, N. (1987). Development of a multi-axes motion control camera system for special effects in animation. *Proceedings of Motortech Japan '87*, pp. I/7/1-14.
5. Kenjo T., Takahashi H., Marushima K., and Cheang-Wee K.S. (1985). Brushless DC drives of a hybrid stepping motor for low-cost applications. *Proceedings of DRIVES/MOTORS/CONTROLS '85*, pp. 183-7.

7 Inverters

Static equipment that is used for converting d.c. power to a.c. power is called an inverter. There are several different types of inverters in terms of the number of phases, the use of solid-state devices, and the output waveforms. We shall look first at the principle of the single-phase inverter and its extension to two-phase inverters. Secondly, discussions will be focused on the principles of three-phase inverters, which are extensively used for variable speed control of induction and synchronous motors, followed by a comparison between current-source and voltage-source inverters. Thirdly, for shaping the output waveforms suited for better drives of motors, a theory for generating PWM signals will be developed on the voltage-source inverter, followed by some examples of hardware and software.

7.1 The H-bridge inverter

The basic principle for generating a single-phase alternating voltage from a d.c. power supply, using a scheme known as the H-bridge circuit, is illus-

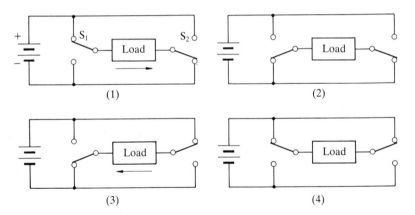

Fig. 7.1. Basic principle of the single-phase bridge inverter using two mechanical switches S_1 and S_2. In state (1), S_1 is on the (+) terminal and S_2 on the (−) terminal, a positive voltage being applied to the load. In state (3), S_1 is on (−) and S_2 on (+), a negative voltage being applied to the load. By causing these two states to alternate repeatedly, a single-phase square-wave voltage is generated. In state (2), both switches are on the (−) side, and in state (4) on the (+) side. In these states, no voltage is applied to the load.

trated in Fig. 7.1. Current is provided to the load from a d.c. power supply via two switches S_1 and S_2. Since each switch has the positive and negative terminals, the combinations of the two switches provides the following four states:

State	S_1	S_2	
1	(+)	(−)	positive voltage at load
2	(−)	(−)	zero voltage at load
3	(−)	(+)	negative voltage at load
4	(+)	(+)	zero voltage at load

When states (1) and (3) are repeated in alternation, a square-wave voltage will be created across the load as illustrated in Fig. 7.2(a). If state (2) or (4), which makes the load potential zero, is used, the waveform of (b) may be obtained. A comparison between these two different waveforms can be made in terms of the harmonic components involved as shown in Table 7.1. The biggest difference here is that the waveform (b) does not involve the third and other $3n^{th}$ harmonic components.

Table 7.1. Ratio of harmonic components involved in the waveforms in Figs. 7.2(a) and (b).

	Square wave (a)	Step wave (b)
Fundamental	1	1˙
3rd harmonic	1/3	0
5th harmonic	1/5	1/5
7th harmonic	1/7	1/7
9th harmonic	1/9	0
11th harmonic	1/11	1/11
13th harmonic	1/13	1/13

If the two switches are to be replaced with transistors, the basic scheme of the single-phase H-bridge inverter will be as shown in Fig. 7.3. Four transistors and also four diodes (cf. Section 2.8.5) are needed. When Tr_1 is ON and Tr_2 is OFF, terminal A is at potential E. When Tr_1 is OFF and Tr_2 is ON, A is at the GND potential. Similarly, when Tr_3 is ON and Tr_4 is OFF, terminal B is at E, and when the switching states are reversed it is on GND.

When the switching state changes from one state to the other between the two paired transistors, both transistors must be on the OFF states for a short time. This is to avoid the possibility of short-circuiting in the transient state in which the two transistors can be simultaneously closing. Hence, switching from the ON state to the OFF must be done in a transistor as quickly as possible, while the switching from the OFF to the ON must be carried out with an appropriate delay and take a definite time.

The switching sequences of the four transistors to produce the waveforms of the square and step waves are given in Fig. 7.2.

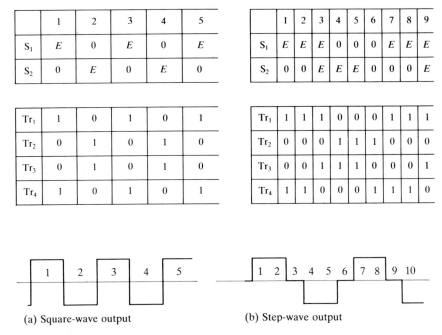

	1	2	3	4	5
S_1	E	0	E	0	E
S_2	0	E	0	E	0

	1	2	3	4	5	6	7	8	9
S_1	E	E	E	0	0	0	E	E	E
S_2	0	0	E	E	E	0	0	0	E

	1	2	3	4	5
Tr_1	1	0	1	0	1
Tr_2	0	1	0	1	0
Tr_3	0	1	0	1	0
Tr_4	1	0	1	0	1

	1	2	3	4	5	6	7	8	9
Tr_1	1	1	1	0	0	0	1	1	1
Tr_2	0	0	0	1	1	1	0	0	0
Tr_3	0	0	1	1	1	0	0	0	1
Tr_4	1	1	0	0	0	1	1	1	0

(a) Square-wave output   (b) Step-wave output

Fig. 7.2. Two examples of switching sequence in the bridge inverter. Method (a) generates a square-wave voltage, and method (b) a step-wave output using all the states shown in the previous figure. The upper tables are for the inverter with mechanical switches S_1 and S_2; the lower tables are for the transistorized inverter.

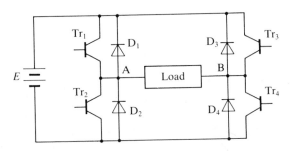

Fig. 7.3. Bridge inverter using four transistors and four diodes.

As an advanced method, if a switching sequence is employed to produce the ON/OFF patterns shown in Fig. 7.4(a) and (b), the voltage waveform appearing across the load will be as in (c). This technique is referred to as 'sinusoidal pulse-width modulation'. The fundamental component in the load voltage is shown by the broken curve in portion (c) and higher harmonic

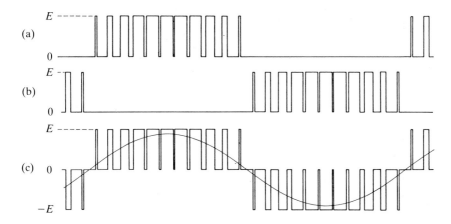

Fig. 7.4. Pulse-width modulation waveform to obtain near-sinusoidal voltage using an H-bridge inverter: (a) voltage at terminal A; (b) voltage at terminal B; (c) the voltage applied to the load. The sinusoidal curve is the fundamental component of the load voltage.

components are few except for the pulsation component. Hence, the current will vary with time as shown in Fig. 7.52, which will be discussed in Sections 7.6 and 7.8.

By combining two H-bridge inverters, as shown in Fig. 7.5, one can get a two-phase inverter that may be used to drive a two-phase induction or synchronous motor.

7.2 Other basic types of single-phase inverters

The H-bridge inverter, which uses four switching devices, is the most orthodox type of inverter. Other types of inverters are illustrated in Fig. 7.6. Type (a) uses two transistors and two d.c. power supplies. Type (b) needs one d.c. power supply, two transistors, and a centre-tapped transformer. This type features electrical isolation between the load and power supply, the load voltage being adjustable by the turn ratio of the transformer. Type (c) employs a large capacitor working as a d.c. battery with a voltage of $E/2$ when the time constant, which is CR for of a resistive load, is much higher than the reciprocal of the switching frequency. Hence, as explained in Fig. 7.7, when Tr_1 is ON and Tr_2 is OFF, the voltage applied to the load is $E - E/2 = E/2$; when the switching states are reversed, it is the capacitor voltage itself ($-E/2$).

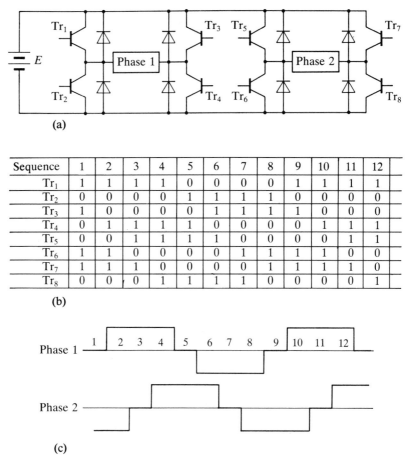

Fig. 7.5. Two-phase inverter and an example of switching sequence, (a) fundamental circuit; (b) an example of switching sequence; (c) waveforms.

7.3 Basic principle of the three-phase inverter

The fundamental principle of obtaining three-phase a.c. currents to be supplied to an a.c. motor is illustrated in Fig. 7.8, in which three mechanical switches are used for commutation of the current supplied from the d.c. source to the stator windings in the proper sequence. A number of sequences for operating these three switches are possible, but there are two fundamental methods that complete one cycle with six switchings; one is known as the 120-degree and the other the 180-degree conduction type.

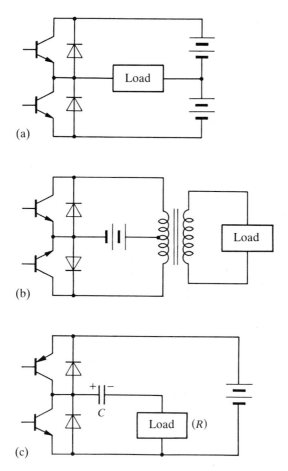

Fig. 7.6. Three types of single-phase inverter that use two power switches: (a) uses two power supplies; (b) employs a centre-tapped transformer; and (c) has a large capacitor.

7.3.1 *120-degree type*

This switching sequence is determined by following the rule that one of the three switches be on the positive terminal, another on the negative terminal, and the last kept open. An example is shown in Fig. 7.9. In this figure the current distribution and resultant magnetic field are illustrated to show how a magnetic field rotates in a motor. A rotor placed in this magnetic field is caused to revolve in the same direction. If the sequence is reversed, the magnetic field and hence the rotor will rotate in the opposite direction, also.

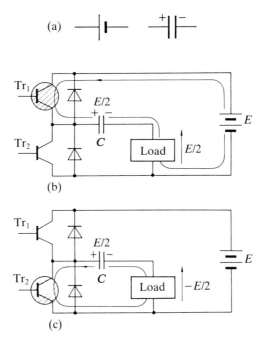

Fig. 7.7. Principle of the single-phase inverter of Fig. 7.6(c). (a) The charged capacitor C works as a battery having a voltage of $E/2$ in the stationary state; (b) hence, when Tr_1 is ON and Tr_2 OFF, the voltage applied to the load is $E/2$; (c) when Tr_1 is OFF and Tr_2 ON, the load voltage is $-E/2$.

It should be noted that in this process a switch is turned to the E and GND side each for a 120° interval in every cycle.

7.3.2 *180-degree type*

Switching for this type is implemented without an OFF period; that is, each switch is always on either the positive or negative terminal, but the situation to be avoided is to have all three on the positive or negative terminal at the same time. The sequence for clockwise rotation is shown in Fig. 7.10. In this figure, the arrangement of the terminals U, V, and W has been rotated 180° for the convenience of discussion of the voltage vector in Section 7.5. For the actual magnetic field distribution inside the motor, refer to Section 1.3.1 and Figs. 1.18 to 1.20.

In either the 120° or 180° type, the magnetic field rotates at 60° intervals. But differences can be seen in the voltage waveforms. Figure 7.11 illustrates the comparison of the two sequence methods when the star-connected resis-

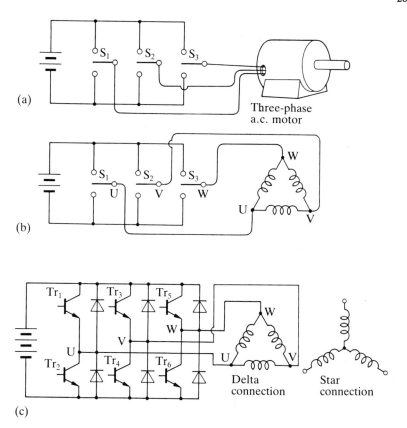

Fig. 7.8. Basic configuration of the three-phase bridge inverter: (a) uses three mechanical switches; when the motor has delta-connected windings, switch terminals are connected as shown in (b); (c) is the transistorized inverter. The motor windings are often connected in the Y (or star) scheme.

tive load is used. The reason that $\frac{1}{3}E$ and $\frac{2}{3}E$ appear in the terminal voltage with respect to the neutral in the 180° operation is explained by Fig. 7.12.

When an inductive load such as an induction motor is driven in the 120° mode, the waveform will deviate from those depicted, because the terminal potentials in the OFF period will be affected by the transient current behaviour. Basically, in the 180° conduction operation the line-to-line waveform is independent of the load characteristics.

In the voltage waveforms across one phase, a difference is seen between the delta and star connections, as shown in Fig. 7.13. However, this difference is not a big problem because there is no difference between the contents of the higher harmonics. Neither includes $3n^{\text{th}}$ harmonics and both contain

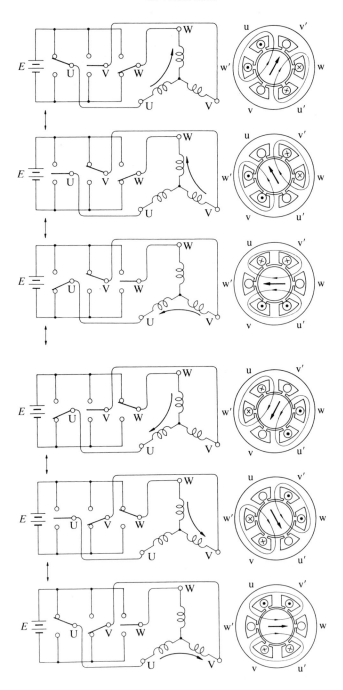

Fig. 7.9. Switching sequence for the 120° conduction inverter.

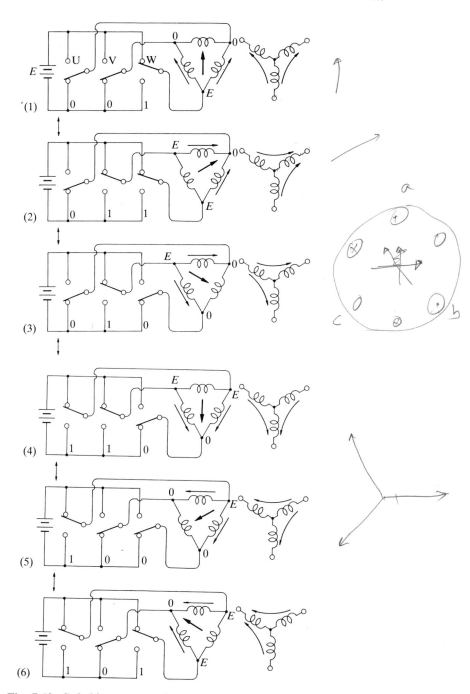

Fig. 7.10. Switching sequence for the 180° conduction inverter.

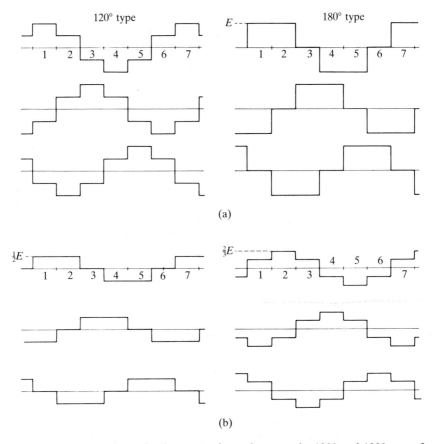

(a)

(b)

Fig. 7.11. Comparison of voltage waveforms between the 120° and 180° types for resistive load: (a) line-to-line voltage; (b) terminal voltage with respect to the neutral point.

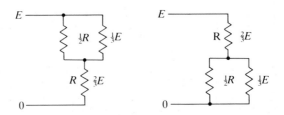

Fig. 7.12. Potentials at the neutral point N.

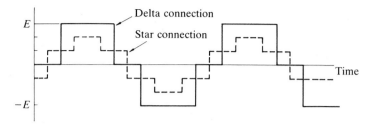

Fig. 7.13. Comparison of the output waveforms between the delta and star connections in the 180° operation.

the same content ratio of the mth harmonic voltage, which is $1/m$. The only difference is seen in the phase angle of the higher harmonics. For details about harmonics readers should consult Reference [1].

Figure 7.14 shows the current waveform flowing in one of the three lines in the 180° operation.

7.4 Voltage-source inverter and current-source inverter

In practical inverters employing solid-state devices, either the d.c. voltage or the d.c. current is unchanged when a commutation takes place. The former is known as the voltage-source inverter, and the latter as the current-source inverter.

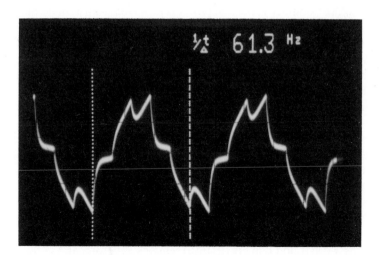

Fig. 7.14. Current waveforms in one of the three lines in the 180° operation.

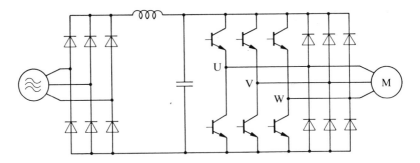

Fig. 7.15. Voltage-source inverter using transistors as the switching elements.

7.4.1 *Voltage-source inverter*

Figure 7.15 shows the circuit of the voltage-source inverter, which has a large capacitor between the d.c. terminals to stabilize the d.c. voltage when a commutation occurs or, in technical terms, to make the output impedance of the power supply the minimum. The 180° mode is employed for the six-step operation of this inverter. However, various sorts of PWM techniques can be employed to improve or control the output waveforms.

A drawback of the voltage-source inverter is that when the load machine works in the generator mode, the electric power generated cannot return to the a.c. power source to save energy. This can be understood using Fig. 7.16. For the sake of simplicity, the d.c. power is provided from a single-phase converter comprising one diode, one reactor, and one capacitor. Now let us suppose that the inverter is operated to produce a 60 Hz three-phase voltage to drive a four-pole induction motor at a speed of 1700 r.p.m. that is lower than the synchronous speed of 1800 r.p.m. In this state the current direction is as shown in (a).

Figure 7.16(c) shows the torque-versus-speed characteristic curves for 60 Hz and 50 Hz operation of this machine. Now the operating point Q is in the first quadrant (positive torque and CW rotation). If the output frequency is suddenly reduced to 50 Hz, the operating point will be shifted to the point Q′ in the fourth quadrant. Note that the speed cannot be changed instantaneously. The torque has become negative, and owing to this the machine has begun to decelerate. The kinetic energy of the rotor and the load is converted to electrical energy that is to be fed back to the power supply. Thus, the motor is now working as a generator. Owing to the large capacitor, the voltage between A and B in Fig. 7.16(b) remains as before, and so the current is reversed to send the generated electric power back to the power source.

The problem now is that because of the diode in the converter, this

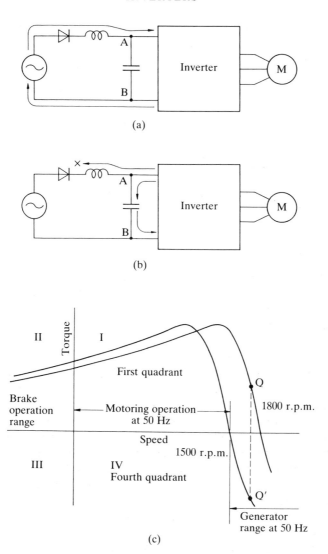

(a)

(b)

(c)

Fig. 7.16. Voltage-source inverter is incapable of feeding back the generated power to the power supply: (a) shows the current direction when the motor is operating in the motoring mode; (b) shows the current in the generator mode, where the diode used in the converter section blocks the feedback current. When much current flows into the capacitor the capacitor voltage suddenly increases and may cause damage to the transistors. (c) Torque-versus-speed characteristics for the 50 Hz and 60 Hz operations, and operating points Q and Q′.

current cannot flow back towards the a.c. supply, and hence it flows into the capacitor to charge it. This results in a sudden increase of the capacitor potential, which often produces damage to the transistors when it exceeds the maximum allowable rating of the collector-to-emitter voltage. To protect the transistors, another circuit must be added, which is used to discharge the electrostatic energy stored in the capacitor and convert it to heat loss when the capacitor voltage exceeds a certain level. The details of this technique were explained in Section 5.7.2 using Fig. 5.27.

It is of course possible to add another circuit that can feed back the generated power to the a.c. supply, but such a scheme can be expensive. Although the overall efficiency of voltage-source inverters is lower than that of current-source inverters described next, they find wide applications in the low- and medium-power areas owing to system simplicity.

It should be noted that the operation of an induction machine in the second quadrant in Fig. 7.16(c) is usually avoided, because the electrical energy supplied from the power source and the mechanical power produced via the machine shaft are both dissipated as heat in the motor, and the temperature rise can be very high.

7.4.2 *Current-source inverter*

Figure 7.17 shows a case in which a large reactance is placed between the d.c. power supply and the switching stage to keep the d.c. current constant or, in technical terms, to provide a high internal impedance for the d.c. power supply. This inverter, which is known as the current-source type, employs thyristors or GTOs as the switching devices that are to be controlled in the 120° mode. The functions of the diodes and capacitors used in this configuration are explained in Section 9.2.2 using Fig. 9.12.

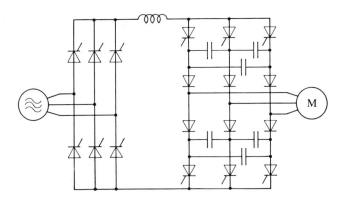

Fig. 7.17. Current-source inverter using thyristors as the switching elements.

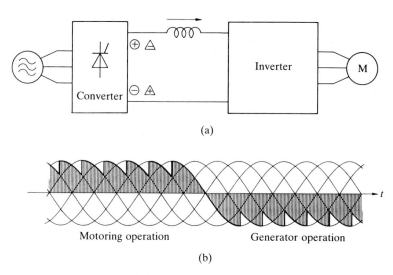

(a)

Motoring operation   Generator operation

(b)

Fig. 7.18. Current-source inverter is capable of feeding back the generated power from the load machine (motor) to the power supply. ⊕ and ⊖ indicate the voltage polarity in the motoring operation, while ▲ and △ represent the polarity in the generator operation. (a) Block diagram, and (b) voltage applied to the load machine.

For control of the output voltage in the current-source inverter, the phase-control technique discussed in Chapter 3 is employed in the converter section, also using thyristors or GTOs. (As we shall soon discuss, the output voltage of the voltage-source inverter is normally adjusted at the inverter section using the PWM technique.) When the motor is brought into the fourth quadrant to work as a generator, electrical power can be fed back to the a.c. power source owing to the combination of a reactance and a converter using power devices whose trigger timing is controllable.

As explained in Fig. 7.18, when the load machine (which is normally a motor) acts as a generator in the fourth quadrant, the current direction is maintained unchanged owing to the reactor, but the polarity of the d.c. voltage can be reversed by control of the gates of the thyristors in the converter section. Thus, the current-source inverter offers good overall efficiency because of the availability of the regenerative brake in decelerating the motor. This type of inverter is used in high-power applications where energy saving is essential. However, since a sophisticated feedback scheme is needed to stabilize the motion of the rotor driven by a current-source inverter, this type is not always suitable for low-power applications.

The capability for both motoring and regenerative operations in either rotational direction is referred to as 'four-quadrant operation'. The coordinates used are the torque and the speed, as shown in Fig. 7.19. The first and

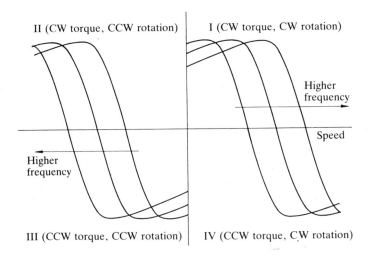

II (CW torque, CCW rotation)   I (CW torque, CW rotation)

Higher
frequency

Speed

Higher
frequency

III (CCW torque, CCW rotation)   IV (CCW torque, CW rotation)

Fig. 7.19. Coverage of the fourth-quadrant operation of an induction motor by a current-source inverter. By changing the output frequency, most of the area can be covered.

fourth quadrant operations are in the CW rotation. Second- and third-quadrant operations are available by changing the switching sequence of the power devices to reverse the direction of the rotating field. It should be noted that in Section 3.5 the definition of the four-quadrant operation of a converter was given in terms of the voltage-versus-current relation.

7.5 Expressing voltage vector rotation in a three-dimensional space

To discuss more sophisticated switching sequences employed in the voltage-source inverter it is convenient to use the concept of the voltage vector and its rotation. Let us consider on the basic inverter of Fig. 7.8. This inverter has three switches S_1, S_2, and S_3, and each switch is either on the positive or the negative terminal. Let the positive state be denoted by 1 and the negative state 0. There may thus be eight different combinations of switching states as follows:

	S_1	S_2	S_3
(1)	0	0	0
(2)	0	0	1
(3)	0	1	1
(4)	0	1	0
(5)	1	1	0

(6)	1	0	0
(7)	1	0	1
(8)	1	1	1

Of these eight combinations, the first and last do not cause a current to flow to the motor, since all the three terminals U, V, W are on the same voltage, GND or E, and hence, the line-to-line voltages are all zero. The other six states appearing in this table can produce voltages to be applied to the motor terminals as already seen in Fig. 7.10, illustrating the sequence of the six-step 180° type inverter.

Now consider a useful method of expressing time variation of the three-phase voltages obtained from an inverter, using three-axis coordinates. Figure 7.20(a) shows this. The three axes U, V, and W are for the state of

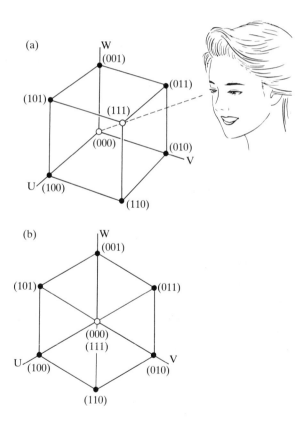

Fig. 7.20. (a) Binary states of the switching elements represented in three-dimensional space, and (b) a hexagon derived from this presentation by viewing along the broken line connecting (000) and (111) in (a).

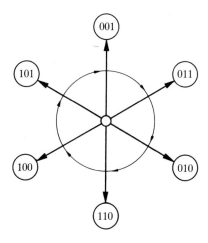

Fig. 7.21. Voltage vector rotation in the 180° conduction-type inverter.

switches S_1, S_2, and S_3 respectively. The eight different states are plotted in the coordinates, and of these (000) and (111) are indicated by open circles (○), and the other six by solid circles (●). These points are the apexes of a cube whose side length is unity. If one views the cube and these apexes from the angle indicated by the broken line connecting (000) and (111) in (a), the image will be a hexagon as shown in (b). In this figure (000) and (111) overlap at the centre.

The arrows produced by connecting the six apexes and the centre are shown in Fig. 7.21, and may be regarded as the magnetic field vectors appearing in Fig. 7.10. Let us here define these as the voltage vectors.

It is seen that in the 180° type inverter operation shown in Fig. 7.10 the voltage vector proceeds in the sequence (001)→(011)→(010)→(110)→(100)→ (101)→(001). If this rotation proceeds in the reverse sequence, the rotor will rotate in the opposite direction. It is evident that the rotation of vectors can begin from any state. Figure 7.22 shows the relation between the (mechanical) switch states, transistor switching states, and terminal voltages when the vector rotates clockwise from (001).

7.5.1 *Hexagon drive*

When the switching frequency is low in the six-step inverter, the rotor movement will display cogging. To smooth the rotor motion, the pulse-width modulation (PWM) technique can be used. The best known PWM is the sinusoidally approximated method, which we shall discuss shortly, but before proceeding to this subject, we shall here note that the six-step method

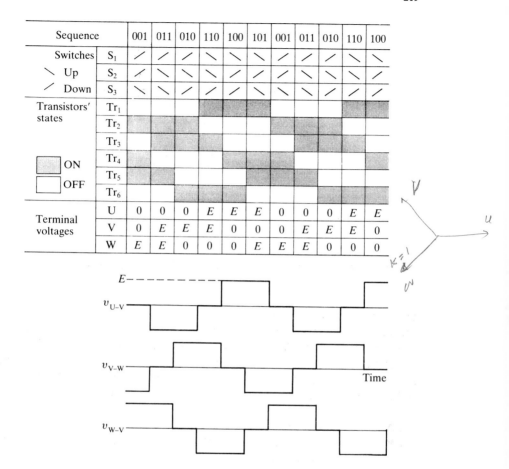

Sequence		001	011	010	110	100	101	001	011	010	110	100
Switches	S₁	╱	╱	╱	╲	╲	╲	╱	╱	╱	╲	╲
╲ Up	S₂	╱	╲	╲	╲	╱	╱	╱	╲	╲	╲	╱
╱ Down	S₃	╲	╲	╱	╱	╱	╲	╲	╲	╱	╱	╱
Transistors' states	Tr₁											
	Tr₂											
	Tr₃											
ON	Tr₄											
OFF	Tr₅											
	Tr₆											
Terminal voltages	U	0	0	0	E	E	E	0	0	0	E	E
	V	0	E	E	E	0	0	0	E	E	E	0
	W	E	E	0	0	0	E	E	E	0	0	0

Fig. 7.22. Relation between voltage waveforms and the voltage vector codes in the 180° mode sequence.

can be improved to produce an effective trapezoidal voltage variation using a PWM technique.

The drawback of the 180° six-step operation is an abrupt transient from a vectorial voltage state, e.g. (100), to another state, e.g. (110). To make this transition gentle, both states may be repeated and the ratio of the period for the state to stay at (100) to the repetition cycle time is gradually increased from 0 to 1. Owing to the inductance in the windings, currents vary more smoothly than the case of six-step operation.

By employing the concept of the PWM here, we can define coordinate values that are other than 0 to 1 and range between these two values. Some of

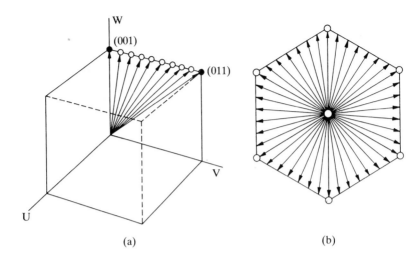

(a)   (b)

Fig. 7.23. Travel of voltage vector in hexagon drive of the voltage-source inverter. (a) Voltage vector locus from (00i̇) to (011); (b) one full cycle.

the coordinate states plotted from calculations based on this idea are shown in Fig. 7.23; (a) shows the travel from the (001) state to the (011) state, and (b) covers one complete cycle. Thus, the voltage vector traces a hexagon in a full cycle. This form of drive of the voltage-source inverter is here termed the 'hexagon drive'. The line-to-line voltage profiles for this drive are illustrated in Fig. 7.24. It is seen that pulse-width modulation is employed in the transit from one state to another. The pulse-width ratio values are indicated by broken lines in (a). Thus, the voltage waveforms are trapezoidal and are closer to a sine wave than the six-step wave shown in Fig. 7.22.

7.5.2 180-degree type PWM

In many applications of inverters to drive an induction or synchronous motor, it is desirable that the ratio of the voltage to the frequency is kept constant. Let us consider this problem on the 180° six-step inverter. A 1/6 cycle is sliced into several pulses, and the pulse frequency is varied with the pulse width kept constant, as illustrated in Fig. 7.25. It is seen that as the pulse frequency decreases the average voltage decreases proportionately. The three-dimensional locus of the voltage vector for this PWM is shown in Fig. 7.26.

It should be noted here that both (000) and (111) are used repeatedly to adjust the output voltage to below the maximum. The waveforms at points U, V, and W and line-to-line voltages are shown in Fig. 7.27.

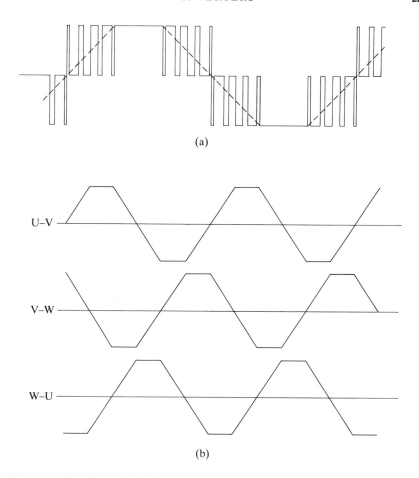

(a)

(b)

Fig. 7.24. Waveforms in the hexagon drives of the voltage-source inverter; (a) is the instantaneous voltage between two lines, and (b) illustrates the relation between the three phases.

7.6 PWM for generating sinusoidal current

When the three-phase voltages that are applied to a three-phase motor vary in a sine-waves shifted 120° relative to each other, the magnetic field rotates in its stator with a constant flux level. In this drive the current waveforms are also sinusoidal. Here we shall discuss methods of producing three-phase sine-wave current using a voltage-source inverter.

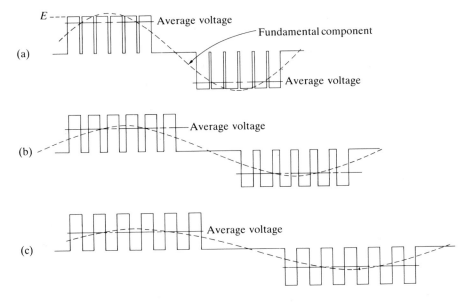

Fig. 7.25. Pulse-width modulation (PWM) of the six-step voltage-source inverter to maintain F/V = constant. (a) High frequency; (b) Medium frequency; (c) Low frequency.

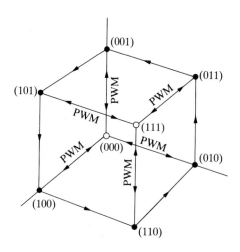

Fig. 7.26. Travel of voltage vector in the 180° PWM operation.

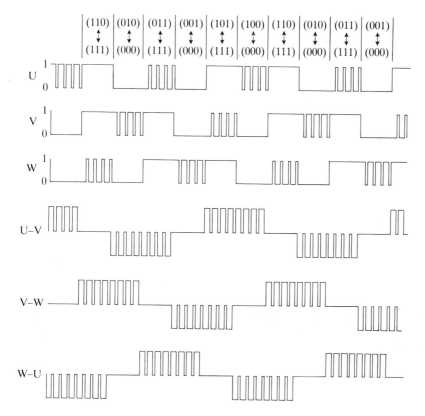

Fig. 7.27. How the voltage vector components (codes) vary with time in the 180° PWM drive of a voltage-source inverter.

7.6.1 *Pulse-width modulation in a sinusoidal pattern*

Let us first consider the case in which each phase is pulse-width modulated at terminals, U, V, and W by the modulation functions:

$$u = \tfrac{1}{2}[m \sin(\omega t) + 1], \tag{7.1a}$$

$$v = \tfrac{1}{2}\left[m \sin\left(\omega t - \frac{2\pi}{3}\right) + 1\right], \tag{7.1b}$$

$$w = \tfrac{1}{2}\left[m \sin\left(\omega t + \frac{2\pi}{3}\right) + 1\right], \tag{7.1c}$$

where m is the modulation factor that can vary from 0 to the maximum of 1. Figure 7.28 illustrates the PWM waveforms when one cycle is divided into 15 pulses for $m = 1$ and for $m = 0.6$.

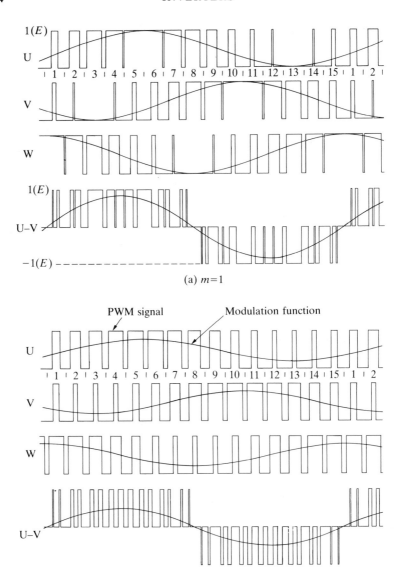

(a) $m=1$

(b) $m=0.6$

Fig. 7.28. Sinusoidal PWM waves with 15 divisions in a cycle: (a) $m=1$; (b) $m=0.6$.

When the d.c. voltage is E, the demodulated line-to-line voltages will be

$$U - V = (\sqrt{3}/2)\, mE \sin\left(\omega t + \frac{\pi}{6}\right), \tag{7.2a}$$

$$V - W = (\sqrt{3}/2)\, mE \sin\left(\omega t + \frac{\pi}{6} - \frac{2\pi}{3}\right), \tag{7.2b}$$

$$W - U = (\sqrt{3}/2)\, mE \sin\left(\omega t + \frac{\pi}{6} + \frac{2\pi}{3}\right), \tag{7.2c}$$

It should be noted that the maximum amplitude, which occurs when $m = 1$, is $(\sqrt{3}/2)E$, and the size of vector locus is obviously smaller than with the hexagon drive as shown in Fig. 7.29(a). Thus, we can conclude that when each phase is modulated in a simple sinusoidal pattern, the output voltage is lower than with the hexagon drive or the plain six-step inverter.

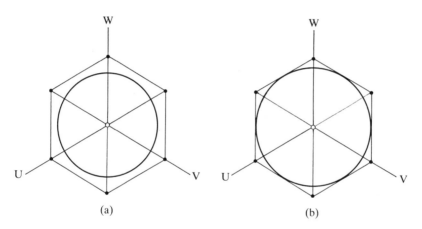

(a)   (b)

Fig. 7.29. Comparison of voltage vector loci; (a) plain sinusoidal PWM with modulation factor $m = 1$; (b) sinusoidal PWM with a superimposed third harmonic with modulation factor $m = 1$.

7.6.2 Improvement by superimposing third harmonics

The maximum circular locus, which contacts the hexagon as shown in Fig. 7.29(b), is obtained when each terminal is modulated in a sinewave with a superimposed third harmonic component whose amplitude equals to 1/6 times the fundamental:

$$u = (1/\sqrt{3})m\left[\sin \omega t + \left(\frac{1}{6}\sin 3\omega t\right)\right] + \frac{1}{2}, \tag{7.3a}$$

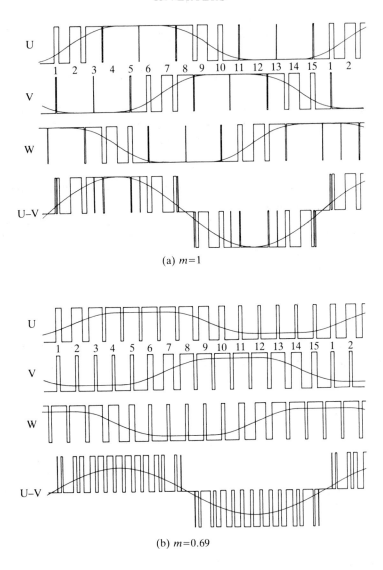

(a) $m=1$

(b) $m=0.69$

Fig. 7.30. Sinusoidal PWM waves having the third-harmonic component; (a) For $m=1$, and (b) $m=0.69$.

$$v = (1/\sqrt{3})m\left[\sin\left(\omega t + \frac{2\pi}{3}\right) + \left(\frac{1}{6}\sin 3\omega t\right)\right] + \frac{1}{2}, \qquad (7.3b)$$

$$w = (1/\sqrt{3})m\left[\sin\left(\omega t - \frac{2\pi}{3}\right) + \left(\frac{1}{6}\sin 3\omega t\right)\right] + \frac{1}{2}. \qquad (7.3c)$$

It should be noted that when the modulation factor $m = 1$, these functions also vary between 0 and 1, and it is obvious that the line-to-line voltages do not include the third harmonics, as expressed by

$$U - V = mE \sin(\omega t + \pi/6). \qquad (7.4)$$

This situation is clearly illustrated in an example shown in Fig. 7.30 in which one cycle is again divided into 15 divisions.

Figure 7.31 shows a comparison of modulation profiles, voltage wave-forms, and the fundamental components between the above-mentioned schemes. It is seen that the six-step 180° drive provides the largest funda-mental component, and the plain sinusoidal PWM the lowest.

7.7 Inverter hardware

Several experimental and sample inverters will be presented here. As there are many similarities between an inverter and a PWM servo-amplifier, circuits shown in Chapter 5 will be used here.

7.7.1 Using combinations of PNP and NPN transistors

Figure 7.32 shows a three-phase inverter that employs a cascade combination of NPN and PNP transistors in the collector-follower scheme discussed with Fig. 2.24 in Section 2.8.4. This is simple and suitable for low-frequency pulse-width modulated operation. In practice, however, power circuits must be protected regarding

(1) over-voltage,
(2) over-current, and
(3) dead time to prevent shorting of two serial arms.

The hardware protectors for over-current and over-voltage are the same as the circuits shown in Figs. 5.29 and 5.30, respectively. However, there is a problem in generating the dead time, which is a short time of some several microseconds to give an OFF signal to both transistors of the Tr_1/Tr_2, Tr_3/Tr_4, and Tr_5/Tr_6 pairs at switching to prevent these transistors from being damaged by short-circuiting during transient period. Since in the circuit of Fig. 7.32 there is no hardware portion that generates the dead

228 INVERTERS

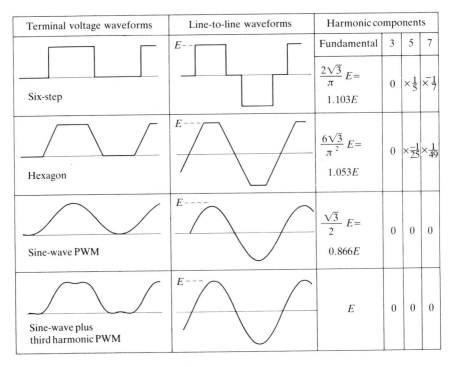

Terminal voltage waveforms	Line-to-line waveforms	Harmonic components			
		Fundamental	3	5	7
Six-step	E	$\dfrac{2\sqrt{3}}{\pi}E=$ 1.103E	0	$\times\frac{1}{5}$	$\times\frac{-1}{7}$
Hexagon	E	$\dfrac{6\sqrt{3}}{\pi^2}E=$ 1.053E	0	$\times\frac{-1}{25}$	$\times\frac{1}{49}$
Sine-wave PWM	E	$\dfrac{\sqrt{3}}{2}E=$ 0.866E	0	0	0
Sine-wave plus third harmonic PWM	E	E	0	0	0

Fig. 7.31. Comparison of waveforms, fundamental and harmonic components between 180° six-step, hexagon, plain sinusoidal, and third-harmonic-added sinusoidal operations.

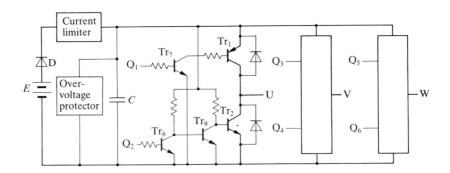

Fig. 7.32. Collector follower type three-phase inverter: This is suited to the low-power experiment with the 180° six-step operation or to low-frequency PWM operations.

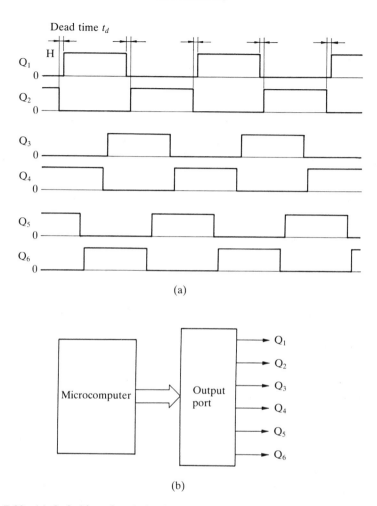

(a)

(b)

Fig. 7.33. (a) Switching signals having dead time t_d; (b) generation of switching signals using a microprocessor.

signal/time, the switching signals generated by a microprocessor must involve dead states for a suitable period, as shown in Fig. 7.33.

7.7.2 Using photocouplers for all NPN transistors

An example of the use of the photocouplers is given as the universal transistor circuit used in the MECHATROLAB that was presented in Chapter 1—see Fig. 1.4(c). This is suitable for operation in a high-frequency PWM mode.

Unlike the example shown in Fig. 7.32, the power and signal stages are isolated by using photocouplers; the switching signals are transmitted via photocouplers. Generation of the dead time is here implemented by hardware using timer ICs for reasons of safety.

7.7.3 MOSFET inverter having two CPUs

The power portion is very similar to the H-bridge converter shown in Fig. 5.27. In most contemporary inverters, at least two microprocessors are used. The example shown in Fig. 7.34 is an inverter system prototyped in the author's laboratory for the drive of MOSFETs in power stages, using an 8749 and 8751 for controlling the switching signals. In this system, however, the PWM signals of a constant frequency of 20 kHz are generated from digital comparators, independently of the output fundamental wave frequencies, which are less than 120 Hz. When the carrier frequency is much higher than the sinusoidal frequencies in this way, asynchronous operation does not represent a serious problem.

In this example, the voltage versus frequency profiles are programmable. The pattern shown in Fig. 7.35 is typical, and a set of different slopes are incorporated in the 8749.

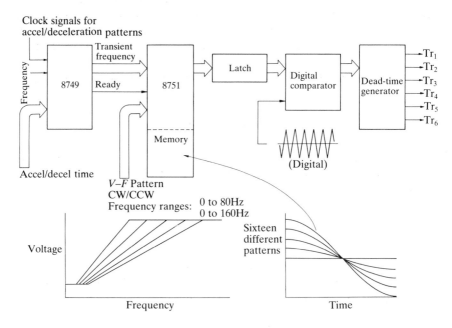

Fig. 7.34. A laboratory-made inverter system designed for MOSFETs as the power devices.

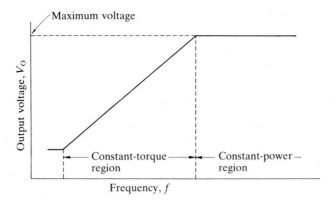

Fig. 7.35. Typical voltage-versus-frequency profile. Before reaching the maximum voltage, constant-torque characteristics will be displayed; after reaching the peak voltage, constant-output characteristics are provided.

Here a power supply of three-phase 200 V, 50/60 Hz is employed, and the maximum output voltage is expected to be 200 V also. As discussed in Section 7.6.1, when plain sinusoidal P W M is employed, the theoretical maximum voltage is 173 V, but in practice it is 160 V or a little higher, owing to voltage drops in the converter and inverter portions. Hence, hexagon operation or the third-harmonic scheme must be employed for getting higher voltages. However, with the third-harmonic method it is difficult to get 200 V owing to the internal losses. The hexagon scheme is therefore the practical choice.

7.8 Some experimental software for the voltage-source inverter

Some examples of experimental software for driving a voltage-source inverter that have been produced for the 8085 or Z80 microprocessor are presented.

7.8.1 *Simple 120-degree operation*

Let us begin with an example of driving a motor in one direction only in 120° operation following the switching sequence in Fig. 7.9. The flowchart is given in Fig. 7.36 and the program list in Table 7.2. Explanations of the program are given in the sequence of the flowchart as follows.

(1) *Initialization.* Register B is used for generating the switching signals, and it is initially loaded with 06H. The meaning of this datum will be

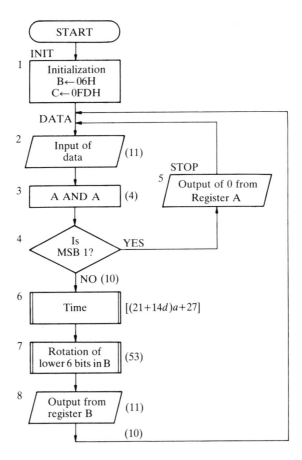

Fig. 7.36. Flowchart for driving an inverter in the 120° conduction mode. Numbers or expressions in ( ) and [ ] are the number of states in the processing, and the total number of states in one cycle in this flowchart is $(21 + 14d)a + 126$, where a and d are as explained in Fig. 7.38.

explained shortly. Register C is used when specifying the output port, and it is initially loaded with 0FDH, which is the address of the output port from the inverter is controlled.

Here follows a detailed explanation of the use of register B. Table 7.3(a) indicates the correspondence between the bit numbers of the output port and the transistor numbers that can produce a proper switching sequence by a simple six-bit rotation of data in a CPU register. Transistors are put in the order 5, 2, 3, 6, 1, 4 from left to right for the order of the bit numbers #5, #4,

Table 7.2. Program listing for generating 120° inverter signals.

```
                    ORG     8500H

8500   06 03   INIT:    LD    B,06H      ;Set initial excitement
8502   0E FD            LD    C,0FDH     ;Set output port

8504   DB FC   DATA:    IN    A,(0FCH)   ;Fetch START/STOP and
                                         ;FREQUENCY data
8506   A7               AND   A          ;Make flags affected
8507   FA 8515          JP    M,STOP     ;If MSB is 1 go to STOP

850A   CD 851B DRIVE:   CALL  TIME       ;Adjust frequency
850D   CD 8526          CALL  ROTATE     ;6-bit rotation
8510   ED 41            OUT   (C),B      ;Step
8512   C3 8504          JP    DATA       ;Go to DATA

8515   AF      STOP:    XOR   A          ;Clear A
8516   D3 FD            OUT   (0FDH),A   ;Turn all transistors off
8518   C3 8504          JP    DATA       ;Go to DATA

               ;****   TIME SUBROUTINE   ****

851B   16 14   TIME:    LD    D,20       ;Load D with 20
851D   15               DEC   D          ;Decrement D until zero
851E   C2 851D          JP    NZ,$-1
8521   3D               DEC   A          ;Decrement A until zero
8522   C2 851B          JP    NZ,TIME
8525   C9               RET              ;Return to main routine

               ;****   ROTATE SUBROUTINE  ****

8526   78      ROTATE:  LD    A,B
8527   C6 E0            ADD   A,0E0H     ;Set carry according to bit #5
8529   17               RLA              ;Rotate left through carry
852A   E6 3F            AND   3FH        ;Clear higher 2 bits
852C   47               LD    B,A
852D   C9               RET              ;Return to main routine

                    END
```

#3, #2, #1, #0 (refer to Fig. 7.8). These switching states proceed from top down, implementing the six-bit rotation from right to left. The highest two bits are not used here. The first state is 000110 or 06H, which turns Tr_1 and Tr_6 on, corresponding to the first state in Fig. 7.9. (refer to Fig. 7.8 again). By rotating these six digits left, the second state 001100 or 0CH is obtained. Another four states can be generated by rotating the data further step by step. The rotation of data is carried out using register B in this program with the higher two bits kept at zero.

Table 7.3. Switching state sequences which are given by a six-bit rotation, and the correspondence between output bit numbers and transistor numbers.

(a) 120° conduction

Transistor number	Bit number of the output port								Hexadecimal code
	7	6	5	4	3	2	1	0	
			5	2	3	6	1	4	
Step 1	0	0	0	0	0	1	1	0	06
2	0	0	0	0	1	1	0	0	0C
3	0	0	0	1	1	0	0	0	18
4	0	0	1	1	0	0	0	0	30
5	0	0	1	0	0	0	0	1	21
6	0	0	0	0	0	0	1	1	03

(b) 180° conduction

Transistor number	Bit number of the output port								Hexadecimal code
	7	6	5	4	3	2	1	0	
			5	2	3	6	1	4	
Step 1	0	0	1	1	0	0	0	1	31
2	0	0	1	0	0	0	1	1	23
3	0	0	0	0	0	1	1	1	07
4	0	0	0	0	1	1	1	0	0E
5	0	0	0	1	1	1	0	0	1C
6	0	0	1	1	1	0	0	0	38

(2) *Input of data.* The command 'Drive or Stop' and operating frequency are given from input port FC as shown in Fig. 7.37. The MSB or the most-significant bit is used for Drive/Stop, and the rest of the bits are for commanding the frequency. The relation between these data and frequency will be discussed later.

(3) *A AND A.* For making the sign flag affected the AND instruction is implemented between the same two data in register A. By this, the data in register A remains unchanged.

(4) *Is MSB 1?* If MSB is 1, the program should be branched to STOP to turn all transistors off and cause the motor to decrease the speed to zero.

(5) *Output zero.* To turn all transistors off, register A is loaded with 0 (zero) and output through the output port FDH. After this execution, the program jumps to DATA to take the command data. Until 'DRIVE' is commanded, the routine will loop.

(6) *Time.* This is the routine to pause to adjust the frequency at a desired value. A detailed flowchart for this part is illustrated in Fig. 7.38. Numbers

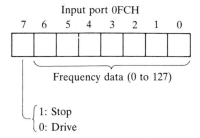

Fig. 7.37. Drive commands given from the input port.

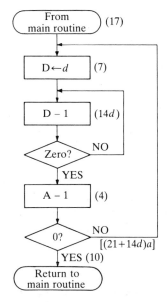

Fig. 7.38. Flowchart of the time routine. The total number of states is $[(21 + 14d)a + 27]$, where d is the datum stored in register D and a is the datum given from the input port.

or expressions in (#) and [#] are the numbers of the states for executing commands. In the overall flowchart of Fig. 7.36, the number of states at each stage is indicated. When the motor is running, the program is executed by following the outer loop repeatedly, and the total number of states is $(21 + 14d)a + 126$ along this loop. One cycle of the output signal is generated by six executions of this loop. Datum d appearing in the flowchart is the datum loaded in register D, being 20 in this program; a is the frequency datum indicated by the lower seven bits in Fig. 7.36.

The output frequency f of the three-phase signal is related to the clock frequency f_c controlling the Z80 microprocessor

$$6[(21 + 14d)a + 126] = f_c/f.$$

Table 7.4 is the data table that generates frequencies from 20 to 120 Hz at 5 Hz intervals.

Table 7.4. Relation between frequency f and data a given from input port for operation of 120° inverter in Table 7.3.

f	a	f	a	f	a
20	110	55	40	90	24
25	88	60	36	95	23
30	73	65	34	100	22
35	63	70	31	105	21
40	55	75	29	110	20
45	49	80	27	115	19
50	44	85	26	120	18

(7) *Rotation of the lower six bits in register B.* Since a rotation of the lower 6 bits cannot be implemented by a simple instruction, a combination of several instructions must be used. The part labelled ROTATE in the program list in Table 7.2 is this routine.

(8) *Output from register B.* By outputting the data from register B to output port FD, one step of the switching state is implemented. After this, the program jumps to DATA.

7.8.2 *Simple 180-degree operation with software dead time*

The switching signals for 180° operation are given in part (b) of Table 7.3, corresponding to the switching sequence in Fig. 7.10. Hence, if initial data 110001B or 31H is loaded into register B, the same program will be usable. However, attention should be paid to one point. If measures for generating dead time discussed in Section 7.7.1 have been implemented in the inverter hardware, there is no problem. If not, dead time must be produced by software to protect the transistors from damage due to short circuiting in transient periods.

(1) *How to produce dead state-codes.* Let us suppose that 07H has been generated, and that the next signal is 0EH. In this transient period, the states of transistors 1, 2, 5, and 6 do not change, and only Tr_3 and Tr_4 change. Before generating 0EH, the AND operation is to be executed between 07H and 0EH and the result, which is the dead-state code, should be generated for

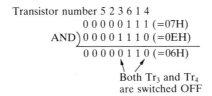

Transistor number 5 2 3 6 1 4

$$\begin{array}{r} 0\,0\,0\,0\,0\,1\,1\,1\ (=07H) \\ AND)\,0\,0\,0\,0\,1\,1\,1\,0\ (=0EH) \\ \hline 0\,0\,0\,0\,0\,1\,1\,0\ (=06H) \end{array}$$

Both Tr$_3$ and Tr$_4$
are switched OFF

Fig. 7.39. How to produce switching data for the dead time.

a short time as the dead time; see Fig. 7.39. It should be noted that in this logical-operational result the switching signals for Tr$_3$ and Tr$_4$ are both 0 (zero). Thus, the dead-state codes can be generated by the AND operation on the states before and after the change.

(2) *Drive instructions.* In the program presented in Table 7.5, the drive instructions are to be given from port FC in the manner shown in Fig. 7.40. The MSB is now used for instruction of rotational direction, and the rest of the bits are for the frequency or the motor speed.

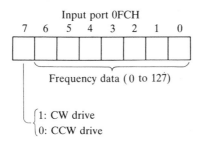

Input port 0FCH

7 6 5 4 3 2 1 0

Frequency data (0 to 127)

1: CW drive
0: CCW drive

Fig. 7.40. Drive commands given from the input port.

(3) *Program.* In this example the registers are used as follows:

Register A: arithmetic and logical operations;
Register B: storing the switching signal that will be generated;
Register C: specifying the output port address;
Register D: storing switching data that have been generated;
Register E: temporary memory of time data;
Register H: storing time adjusting data.

The overall flowchart and details of the 'adjust time' part are presented in Fig. 7.41. The following are the detailed explanation of the program.

Table 7.5. Program listing for generating 180° inverter signals.

```
                              ORG     8600H

8600    06 31    INIT:    LD     B,31H      ;Set initial switching code
8602    0E FD             LD     C,0FDH     ;Set output port
8604    16 00             LD     D,0        ;Load D with zero
8606    ED 41             OUT    (C),B

8608    DB FC    INPUT:   IN     A,(0FCH)   ;Fetch freqency/direction data
860A    5F                LD     E,A        ;and store it in E
860B    50                LD     D,B

860C    A7       DIRECT:  AND    A          ;Direction discrimination
860D    FA 861D           JP     M,CW       ;If MSB is "H", then go to CW

8610    78       CCW:     LD     A,B        ;CCW rotation
8611    E6 20             AND    20H
8613    07                RLCA
8614    07                RLCA
8615    80                ADD    A,B
8616    07                RLCA
8617    E6 3F             AND    3FH
8619    47                LD     B,A
861A    C3 862A           JP     TIME

861D    78       CW:      LD     A,B        ;CW rotation
861E    E6 01             AND    1
8620    0F                RRCA
8621    0F                RRCA
8622    80                ADD    A,B
8623    0F                RRCA
8624    E6 3F             AND    3FH
8626    47                LD     B,A
8627    C3 862A           JP     TIME

862A    7B       TIME:    LD     A,E
862B    26 0D             LD     H,13
862D    E6 7F    LP1:     AND    7FH
862F    5F                LD     E,A
8630    00       LP2:     NOP
8631    E6 7F             AND    7FH
8633    1D                DEC    E
8634    C2 8630           JP     NZ,LP2
8637    25                DEC    H
8638    C2 862D           JP     NZ,LP1
863B    7A       DEAD:    LD     A,D        ;
863C    A0                AND    B
863D    D3 FC             OUT    (0FCH),A   ;
863F    00                NOP
8640    00                NOP
8641    00                NOP
8642    00                NOP
8643    00                NOP
8644    00                NOP
8645    00                NOP

8646    ED 41    DRIVE:   OUT    (C),B      ;Output dead state
8648    C3 8608           JP     INPUT

                              END
```

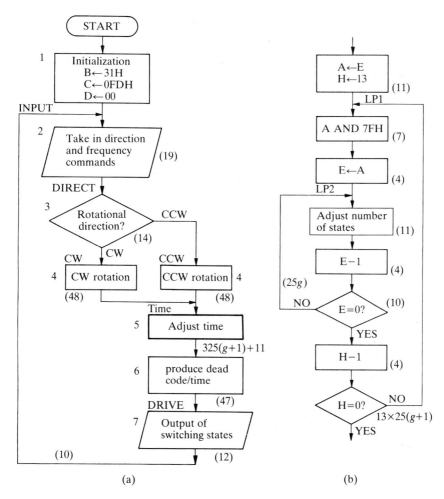

(a)   (b)

Fig. 7.41. Flowchart for generating the 180° conduction signals: (a) overall flow-chart, and (b) details of the 'adjust time' part. The numerals in ( ) are the numbers of states involved in executing commands in a Z80.

1. *Initialization.* Registers B, C, and D are set as follows:

Register B: at 31H, which is the first state in Table 7.3(b);
Register C: at 0FDH, which is the output port address;
Register D: at 0.

The first switching state is sent out from register B to the output port whose address is 0FDH.

2. *Take in direction and frequency commands.* The direction and frequency commands are taken into register A from the input port whose

address is FCH, and also stored in register E. After this the switching state in register B is loaded into register D for the preparation of producing the dead-state code.

3. *Judge rotational direction.* The direction command is in register A. By taking AND with the same datum in register A, the flag register is affected without changing the data itself. If the MSB is 1, the program jumps to CW; if it is 0 to CCW.

4. *CCW/CW rotation.* The lower six bits are to be rotated. However, we do not have a single command for the six-bit rotation and a combination of several commands must be used. In these examples, the higher two bits are set to zero. The statement JP TIME in the CCW routine is only for adjusting time.

5. *Adjust time.* Details of this part are seen in Fig. 7.41(b). The command datum is again transferred to register A, and register H is loaded with 13. The MSB datum in register A is cleared before the datum is reloaded in E. By decounting the data in registers E and H until both are zero, a certain time is expended.

6. *Produce dead code/time.* Now register B stores the new excited state, and D the old state. Using register A, the AND operation between these data is executed to produce the dead code to be sent out to the output port. Dead time is created by seven executions of the NOP (no operation) instruction and one OUT instruction. Its total number of states is 40, and a time of 10 μs is involved if the clock frequency is 4 MHz.

7. *Output the switching states.* To apply the proper potential to the motor, the datum in register B is sent out to the output port. After this the program jumps to INPUT.

Computation of frequency. The frequency can be computed from the time needed to complete one cycle of the loop indicated by INPUT in Fig. 7.41. If this is denoted by T, we have

$$T = \frac{S}{f_c} \text{ (s)},\tag{7.5}$$

where f_c = clock frequency, which is 4 MHz in our case, and
S = total number of states along the loop.

When the main routine is executed six times, the initial exciting state is recovered, and one cycle is completed. The fundamental frequency f_m applied to the motor is, therefore, given by

$$f_m = \frac{1}{6T} = \frac{f_c}{6S} \text{ (Hz)}.\tag{7.6}$$

The total number of states in the loop is

$$S = 325g + 486,\tag{7.7}$$

where g is the time data taken through the input port.
We have

$$f_m = (4/6) \times 10^6/(325g + 486),\qquad(7.8)$$

from which we obtain for the time data

$$g = [(2/3) \times 10^6/f_m - 486]/325.\qquad(7.9)$$

The relation between f_m and g is given in Table 7.6.

Table 7.6. Relation between frequency f and time data given from the input port for the operation of the program in Table 7.5

Frequency f_m (Hz)	Datum g	Time data	
		Integer approx.	Hexadecimal
20	101.1	101	65
30	66.9	67	43
40	49.8	50	32
50	39.5	40	28
60	32.7	33	21
70	27.8	28	1C
80	24.1	24	18
90	21.3	21	15
100	19.0	19	13
110	17.2	17	11
120	15.6	16	10
130	14.3	14	0E
140	13.2	13	0D

7.8.3 Pulse-width modulated 180-degree inverter

The next example is a variable-frequency, variable-voltage inverter, using the simple PWM technique explained in Section 7.5.2. The program flowchart and listing are shown in Fig. 7.42 and Table 7.7, respectively. Registers are used for the following purposes:

Register A: arithmetic and logic operations and as the port for transferring data between data area and I/O ports;

Register B: temporary memory of the switching states;

Register C: memory for the switching states producing zero potential;

Register D: specifying the number of pulses per 1/6 cycle;

Register E: counting pulses generated;

Pair register HL: specifying the memory address storing time data;

Port FC: input port through which frequency data are instructed;

Port FD: output port that sends out switching signals to the transistors.

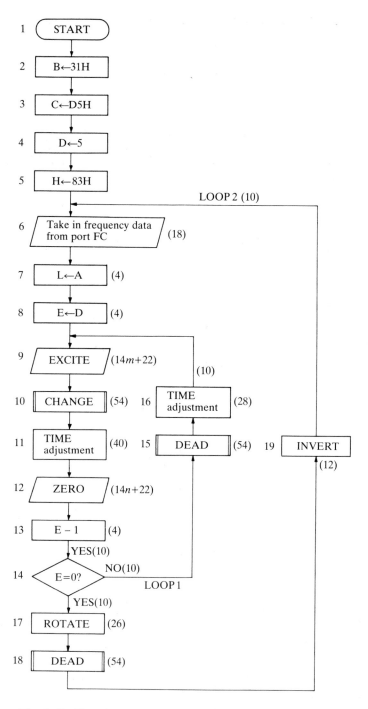

Fig. 7.42. Flowchart for generating P W M signals for an inverter.

Table 7.7. Program listing for generating PWM inverter signals.

```
                    ;****    PWM INVERTER PROGRAM    ****

                            ORG     8200H

00FD                DRIVE   EQU     0FDH        ;Output address
00FC                FRQCY   EQU     0FCH        ;Input  address

8200    06 31       INIT:   LD      B,31H       ;Set switching code
8202    0E D5               LD      C,0D5H      ;Set zero voltage
8204    16 04               LD      D,4         ;Set pulse number
8206    26 83               LD      H,83H       ;Set higher byte address

8208    DB FC       LOOP2:  IN      A,(FRQCY)   ;Fetch frequency data
820A    E6 0F               AND     0FH         ;Clear higher 4 bits
820C    6F                  LD      L,A         ;and store it in L
820D    5A                  LD      E,D         ;Store pulse number in E

820E    78          EXCITE: LD      A,B         ;Move switching code to B
820F    D3 FD               OUT     (DRIVE),A   ;Initial Excitement
8211    3E 64               LD      A,100       ;Pulse width
8213    3D                  DEC     A
8214    C2 8213             JP      NZ,$-1

8217    CD 8250     CHANGE: CALL    DEAD        ;Dead time
821A    00          TIME1:  NOP                 ;Time adjustment
821B    00                  NOP
821C    00                  NOP
821D    00                  NOP
821E    00                  NOP
821F    00                  NOP
8220    00                  NOP
8221    00                  NOP
8222    00                  NOP
8223    00                  NOP

8224    79          ZERO:   LD      A,C         ;Zero voltage
8225    D3 FD               OUT     (DRIVE),A   ;
8227    7E                  LD      A,(HL)      ;Spend time
8228    3D                  DEC     A           ;by decrementing A
8229    C2 8226             JP      NZ,$-1      ;till zero

822C    1D          COUNT:  DEC     E           ;Decount pulse
822D    C2 823E             JP      NZ,LOOP1    ;If not zero, go to LOOP1

8230    78          ROTATE: LD      A,B         ;6-bit rotation
8231    C6 E0               ADD     A,0E0H
8233    17                  RLA
8234    E6 3F               AND     03FH
8236    47                  LD      B,A
8237    CD 8250             CALL    DEAD        ;Dead time
823A    79          INVERT: LD      A,C         ;Move zero-voltage code to A
823B    2F                  CPL                 ;Invert zero-voltage code
823C    4F                  LD      C,A         ;Move data from A to C
823D    C3 8208             JP      LOOP2       ;Go to LOOP2

8240    CD 8250     LOOP1:  CALL    DEAD        ;Dead time
8243    C6 00       TIME2:  ADD     A,0         ;Time adjustment
8245    C6 00               ADD     A,0
8247    C6 00               ADD     A,0
8249    C6 00               ADD     A,0
824B    C3 820E             JP      EXCITE      ;Go to EXCITE
```

Table 7.7. *continued*

```
                              ;****   SUBROUTINE DEAD TIME    ****

                              ORG    8250H

8250    79          DEAD:     LD     A,C
8251    A0                     AND    B            ;Make dead-state code
8252    D3 FD                  OUT    (DRIVE),A
8254    00                     NOP                 ;Time adjustment
8255    00                     NOP
8256    C9                     RET

                              ;****   FREQUENCY DATA    ****

                              ORG    8300H

8300    DE B4 78 92           DEFB   222,180,120,146,52,65,98
8304    34 41 62
8307    50 00 00 07           DEFB   80,0,0,7,1,41,31,14,22
830B    01 29 1F 0E
830F    16                                         ;Data in 8308H and 8309H
                                                    can be any
                              END
```

The sequence of the switching signals used in this example is depicted in Fig. 7.27. As shown in Table 7.8, the switching states for excitation and zero voltage are generated in turn, and are sent out from the output port, and the time spent for the zero-voltage states is variable.

Explanation of the software is given in the order of the numbers in the flowchart:

(1) Program starts from 8200H.

(2) The first switching state is set in register B.

(3) Register C is loaded with 0D5H, which is the second switching datum in Table 7.8. When this is sent to the inverter, Tr_1, Tr_3 and Tr_5 are OFF and Tr_2, Tr_4 and Tr_6 are ON.

(4) Four pulses per 1/6 cycle are generated.

(5) The higher address 83H for the time data is specified.

(6) *LOOP 2*. Frequency instruction data is given from port FC. For simplicity, the frequencies can be instructed between 35 Hz and 100 Hz at 5 Hz intervals. The lower four bits are used for specifying the frequency as will be explained later using Table 7.10. The higher four bits in register A are cleared by the instruction AND 0FH.

(7) The datum in register A is duplicated in register L.

(8) Data in register D, which gives the number of pulses in a 1/6 cycle, is duplicated in register E to be used as a counter.

(9) *EXCITE*. Switching data stored in register B is sent out to the

Table 7.8. Switching sequence for a PWM drive.

		Transistor number		1	4	5	2	3	6	Hexadecimal
		Bit number	7	6	5	4	3	2	1	0	
	1	0	0	1	1	0	0	0	1	31
		1	1	0	1	0	1	0	1	D5
	2	0	0	1	0	0	0	1	1	23
		0	0	1	0	1	0	1	0	2A
Step sequence	3	0	0	0	0	0	1	1	1	07
		1	1	0	1	0	1	0	1	D5
	4	0	0	0	0	1	1	1	0	0E
		0	0	1	0	1	0	1	0	2A
	5	0	0	0	1	1	1	0	0	1C
		1	1	0	1	0	1	0	1	D5
	6	0	0	1	1	1	0	0	0	38
		0	0	1	0	1	0	1	0	2A

inverter through output port FD. After this, a time is spent by setting 100 in register A and decrementing it until zero.

(10) *CHANGE*. Subroutine DEAD is called. Data in register C, which is a switching state for zero voltage, is duplicated in register A and it is ANDed with the data in register B to produce the dead-state code. This is sent out to the transistors.

(11) *TIME* 1. to produce a dead time of about 18 μs, → eight-times NOP (no operation).

(12) *ZERO*. Zero voltage is produced by

$$\text{LD} \quad \text{A,C}$$
$$\text{OUT} \quad \text{(DRIVE), A.}$$

The time data stored in the memory area specified by pair register HL is transferred to register A, and it is decremented until zero to expend a certain time.

(13) Register E is decremented.

(14) If $E \neq 0$ a 1/6 cycle has not yet completed, and so the program branches to produce more pulses.

(15) *DEAD*. A dead state code is to be produced in the transit between zero voltage and excitement.

(16) *TIME* 2. A dead time of about 18 μs is produced.

(17) *ROTATE*. The lower six bits in register B are rotated using register A and the carry bit.

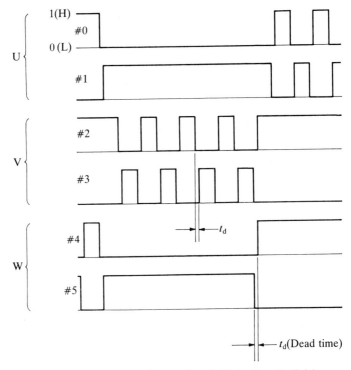

Fig. 7.43. Output waveforms of each bit in the nth division.

(18) The dead state signal is produced and dead time is expended in the transient between zero voltage and excitation. See Fig. 7.43.

(19) *INVERT*. Data in register C is inverted using register A. A new frequency datum is then taken into the CPU and the same routine is repeated.

Time calculations. In the looping part of the flowchart of Fig. 7.42, the number of states in each stage is indicated. Table 7.9 is the calculation table for the number of states. The total number of states is $[14(m + n) + 244]d + 36$.

The output frequency f and clock frequency f_c are related

$$f = \frac{f_c/6}{[14(m + n) + 244]d + 36}. \tag{7.10}$$

From this we get

$$n = [((f_c/6f) - 36)/d - 244]/14 - m. \tag{7.11}$$

Putting $m = 100$, $f_c = 4 \times 10^6$ Hz and $d = 4$, the time data n are given as a function of f as in Table 7.10.

Table 7.9.   Calculation table of the number of states in one loop.

	Label	Instructions		Number of states	Details and total
	LOOP2:	IN	A,		
			(FRQCY)	11	
		AND	0FH	7	
		LD	L,A	4	
		LD	E,A	4	
For	ROTATE:	LD	A,B	4 ⎫	
inner		ADD	A,0EH	7 ⎪	
loop		RLA		4 ⎬ 26	
		AND	03H	7 ⎪	
		LD	B,A	4 ⎭	
		CALL	DEAD	54*	
	INVERT:	LD	A,C	4 ⎫	
		CPL		4 ⎬ 12	
		LD	C,A	4 ⎭	Partial total
		JP	LOOP2	10	= 136
	EXCITE:	LD	A,B	4 ⎫	
		OUT	(DRIVE),A	11 ⎪	(14m+22)d
		LD	A,100	7 ⎬	where m=constant
		DEC	A	4m ⎪	f=number of pulses
		JP	NZ,$-1	10m ⎭	per 1/6 cycle
	CHANGE:	CALL	DEAD	54* ⎫	
		NOP	10 times	40 ⎭ 94d	
	ZERO:	LD	A,C	4 ⎫	
For		OUT	(DRIVE),A	11 ⎪	(14n+22)d
outer		LD	A,(HL)	7 ⎬	where n=time data
loop		DEC	A	4n ⎪	
		JP	NZ,$-1	10n ⎭	
	COUNT:	DEC	E	4 ⎫	
		JP	NZ,		⎬ 14d	
			LOOP1	10 ⎭	
	LOOP1:	CALL	DEAD	54* ⎫	
		AND	A,0		⎪	
			4 times	28 ⎬ 92(d-1)	
		JP	EXCITE	10 ⎭	
				Total = [14(m+n)+244]d+36	
	DEAD*	CALL	DEAD	17 ⎫	
	rotine	LD	A,C	4 ⎪	
	(details)	AND	B	4 ⎬ 54	
		OUT	(DRIVE),A	11 ⎪	
		NOP	2 times	8 ⎪	
		RET		10 ⎭	

Table 7.10.   Relation between output frequency, time data and memory addresses.

Frequency	Lower 4 bits	Memory addresses	Time datum n decimal	Hexadecimal
35	0000	8300	222	DE
40	0001	8301	180	B4
45	0011	8303	146	92
50	0010	8302	120	78
55	0110	8306	98	62
60	0111	8307	80	50
65	0101	8305	65	41
70	0100	8304	52	34
75	1100	830C	41	29
80	1101	830D	31	1F
85	1111	830F	22	16
90	1110	830E	14	0E
95	1010	830A	7	07
100	1011	830B	1	01

Note:
(1) Time index data are computed with $m = 100$, $d = 4$.
(2) Data in the second column are placed in the order of the Gray binary codes. If the datum is denoted by a, the frequency f is given by $f = 5(a + 7)$.
(3) Data for the second column are given from port FCH.

As already stated, the frequency datum is instructed from the lower four bits of the input port. For example, when a frequency of 75 Hz is needed, the code is 1100, and the higher four bits do not affect the frequency.

7.8.4 *Third-harmonic-added sinusoidally PWM inverter*

An example of experimental program for generating the third-harmonic-added sinusoidal PWM signals discussed in Section 7.6.2 is given in Fig. 7.44 in the form of flowchart. The program listing consisting of the main routine and data area for the PWM signals is given in Table 7.11.

This example has 27 divisions in a cycle, and the details of one division are illustrated in Fig. 7.45. It is seen that one division has seven sub-divisions. However, since the last state of 2AH is the same as the first state of the next division, one division can be divided into six sub-divisions in the algorithm, and therefore there are 162 sub-divisions all in a cycle. In this example the dead-state codes and time must be produced by hardware.

The data area from 8101H to 81A4H is for the switching states, and that from 8201H to 82A4H is for the time-data generating pulse width. When the clock frequency is 4 MHz, the fundamental frequency is 70 Hz. These data have been computed by setting the modulation factor at $m = 0.9873$.

The use of registers in this program is as follows:

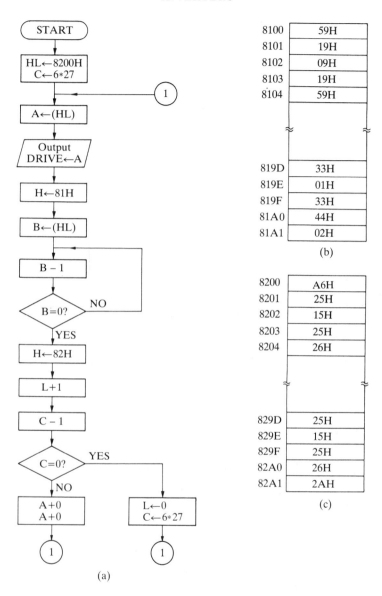

8100	59H
8101	19H
8102	09H
8103	19H
8104	59H
≈	≈
819D	33H
819E	01H
819F	33H
81A0	44H
81A1	02H

(b)

8200	A6H
8201	25H
8202	15H
8203	25H
8204	26H
≈	≈
829D	25H
829E	15H
829F	25H
82A0	26H
82A1	2AH

(c)

(a)

Fig. 7.44. (a) Flowchart for generating switching codes for the sinusoidal P W M drive, one cycle being chopped in 35 divisions. (b) Memory map for time data. (c) Memory map for switching codes.

Table 7.11. Program listing for generating switching signals for the third-harmonics superimposed sinusoidal P W M drive.

```
            ORG     8000H

DRIVE   EQU     0FDH        ;Define output port
BYTE    EQU     162         ;Number of bytes

JOB:    LD      HL,8200H    ;Load HL with 8200H
        LD      C,BYTE      ;Load C with BYTE
LOOP1:  LD      A,(HL)      ;Transfer data from (HL) to A
        OUT     (DRIVE),A   ;Output to DRIVE
        LD      H,81H       ;Load H with 81H
        LD      B,(HL)      ;Tranfer data from (HL) to B
LOOP2:  DEC     B           ;Decrement B
        JP      NZ,LOOP2    ;till zero
        LD      H,82H       ;Load H with 81H
        INC     L           ;Increment L
        DEC     C           ;Decrement C
        JP      Z,RESET     ;If zero go to RESET
        ADD     A,0         ;Adjust time using this command
        LD      A,A         ;Adjust time using this command
        JP      LOOP1       ;Go to LOOP1
RESET:  LD      L,0         ;Clear L
        LD      C,BYTE      ;Reload C with BYTE
        JP      LOOP1       ;Go to LOOP1

;***** TIME DATA *****
        ORG     8100H

TTADA:  DB      59H,19H,09H,19H,59H,04H
        DB      66H,03H,19H,03H,66H,04H
        DB      55H,1EH,07H,1EH,55H,02H
        DB      3FH,38H,01H,38H,3FH,02H
        DB      27H,4EH,03H,4EH,27H,0BH
        DB      0BH,61H,05H,61H,0BH,11H
        DB      10H,5EH,05H,5EH,10H,09H
        DB      2BH,4BH,01H,4BH,2BH,02H
        DB      44H,33H,01H,33H,44H,02H
        DB      59H,19H,09H,19H,59H,04H
        DB      66H,03H,19H,03H,66H,04H
        DB      55H,1EH,07H,1EH,55H,02H
        DB      3FH,38H,01H,38H,3FH,02H
        DB      27H,4EH,03H,4EH,27H,0BH
        DB      0BH,61H,05H,61H,0BH,11H
        DB      10H,5EH,05H,5EH,10H,09H
        DB      2BH,4BH,01H,4BH,2BH,02H
        DB      44H,33H,01H,33H,44H,02H
        DB      59H,19H,09H,19H,59H,04H
        DB      66H,03H,19H,03H,66H,04H
        DB      55H,1EH,07H,1EH,55H,02H
        DB      3FH,38H,01H,38H,3FH,02H
        DB      27H,4EH,03H,4EH,27H,0BH
        DB      0BH,61H,05H,61H,0BH,11H
```

Table 7.11. *continued*

```
          DB      10H,5EH,05H,5EH,10H,09H
          DB      2BH,4BH,01H,4BH,2BH,02H
          DB      44H,33H,01H,33H,44H,02H

          ;***** SWITCHING DATA *****
          ORG     8200H

SDATA:    DB      0A6H,25H,15H,25H,26H,2AH
          DB      26H,16H,15H,16H,26H,2AH
          DB      26H,16H,15H,16H,26H,2AH
          DB      26H,16H,15H,16H,26H,2AH
          DB      26H,16H,15H,16H,26H,2AH
          DB      26H,16H,15H,26H,26H,2AH
          DB      1AH,16H,15H,16H,1AH,2AH
          DB      1AH,16H,15H,16H,1AH,2AH
          DB      1AH,16H,15H,16H,1AH,2AH
          DB      1AH,16H,15H,16H,1AH,2AH
          DB      1AH,19H,15H,19H,1AH,2AH
          DB      1AH,19H,15H,19H,1AH,2AH
          DB      1AH,19H,15H,19H,1AH,2AH
          DB      1AH,19H,15H,19H,1AH,2AH
          DB      1AH,19H,15H,19H,1AH,2AH
          DB      29H,19H,15H,19H,29H,2AH
          DB      29H,19H,15H,19H,29H,2AH
          DB      29H,19H,15H,19H,29H,2AH
          DB      29H,19H,15H,19H,29H,2AH
          DB      29H,25H,15H,25H,29H,2AH
          DB      29H,25H,15H,25H,29H,2AH
          DB      29H,25H,15H,25H,29H,2AH
          DB      29H,25H,15H,25H,29H,2AH
          DB      29H,25H,15H,25H,29H,2AH
          DB      26H,25H,15H,25H,26H,2AH
          DB      26H,25H,15H,25H,26H,2AH
          DB      26H,25H,15H,25H,26H,2AH
COUNT:    DB      162

          END
```

Register A: transferring the switching and time data between the memory and the output port;

Register B: the time counter for generating pulse width;

Register C: counting the serial sub-division number in a cycle;

Register H: specifying the higher address of the data area;

Register L: specifying the lower address of the data area.

The correspondence between the transistor numbers and bit numbers in this program is as follows:

```
Bit number:         7 6 5 4 3 2 1 0
Transistor number:  / / 6 5 4 3 2 1
```

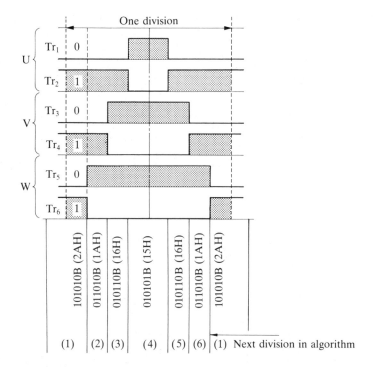

Fig. 7.45. Examples of switching signal waveforms in one division which consists of serial seven states. In the algorithm, one division comprises six sub-divisions, because the last 2A H is the same as the first sub-division state in the next division.

In this program, we have the following relation between the fundamental frequency f, clock frequency f_c, and the total-time data h of the seven sub-divisions in a division:

$$\frac{f_c}{f} = 15(539 + 14h) \tag{7.12}$$

If one wants to vary the frequency, one must change the time data according to this formula.

Figure 7.46 shows the current waveforms seen in three different rotors (squirrel-cage, solid-steel and semihard-steel (or hysteresis) rotors provided in MECHATRO LAB) driven in stator A by its four-pole windings. It is seen that the current in the solid-steel eddy-current motor and the hysteresis motor has less ripple than that in the squirrel-cage motor. Though not seen here, differences between lock-load and no-load waveforms are small in the solid-steel eddy-current motor and the hysteresis motor. When the number of

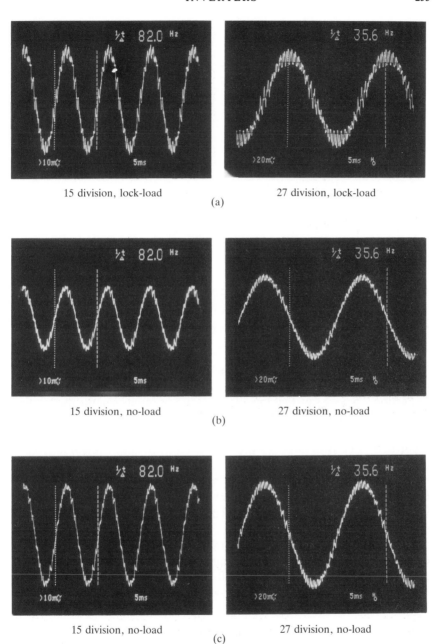

15 division, lock-load   27 division, lock-load

(a)

15 division, no-load   27 division, no-load

(b)

15 division, no-load   27 division, no-load

(c)

Fig. 7.46. Current waveforms when 15-division and 27-division sine-wave PWM voltages are applied to the three-phase windings of (a) a squirrel-cage induction motor, (b) a solid-steel eddy-current motor, and (c) a hysteresis motor.

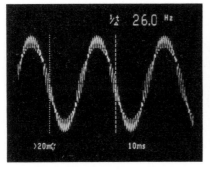 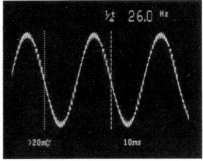

(a) Solid-steel motor at no-load (b) Squirrel-cage motor at lock-load

Fig. 7.47. Current waveform observed in the solid-steel motor and squirrel-cage motor in 39-division 26 Hz operation.

divisions is increased, the current ripples decrease, as shown in Fig. 7.47, which is observed at the 39-division standstill operation of the solid-steel motor.

7.9 An idea for sinusoidal PWM based on vector rotation

In Section 7.6.2 we saw that inclusion of third harmonics in the modulation curve for each phase is useful for utilizing the maximum d.c. voltage. Further utilizing $3n^{th}$ harmonics is an interesting problem and several ideas have been reported.[2-4] Let us discuss this problem in terms of employing a microprocessor in generating the PWM signals for this purpose. For this discussion we shall utilize the concept of the voltage vector and its rotation as introduced in Section 7.5. To make the idea clear, we choose a relatively simple example in which the PWM signals are generated based on the voltage vectors travelling on the cubic surface.

See Fig. 7.48 to define a vector A on the plane surrounded by the three lines connecting three apexes (000), (001), and (101). Let the time for generating a pulse for this vector be T, and this be divided into three components t_0, t_1, and t_2 that the switching state accommodates at the three points as follows:

$$t_0 \quad \text{at (000);}$$
$$t_1 \quad \text{at (001);}$$
$$t_2 \quad \text{at (101).}$$

The geometrical principle for such a time division is given in Fig. 7.48 and the algebraic expressions are

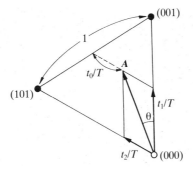

Fig. 7.48. Definition of a voltage vector on a cubic surface, and how to determine time to be spent at each apex.

$$t_1 = qT\left(\frac{2}{\sqrt{3}}\right) \sin\left(\frac{\pi}{3} - \Theta\right), \qquad (7.13)$$

$$t_2 = qT\left(\frac{2}{\sqrt{3}}\right) \sin\Theta, \qquad (7.14)$$

$$t_0 = T - t_1 - t_2 \qquad (7.15)$$

where θ is the angle of inclination of the vector from the W axis, and its maximum value is 60°; q is a quantity similar to the modulation factor, its maximum value being $\sqrt{3}/2 = 0.866$ so that the vector travels on the maximum circle.

As shown in Fig. 7.49, let us suppose that the vector is travelling on the six planes in sequence as follows:

Three points forming a plane

1st plane	(000)	(001)	(101)
2nd plane	(111)	(101)	(100)
3rd plane	(000)	(100)	(110)
4th plane	(111)	(110)	(010)
5th plane	(000)	(010)	(011)
6th plane	(111)	(011)	(001)

Obviously, the vector locus is not continuous when seen from an arbitary angle; it is only a group of six arcs disconnected from each other. The locus is a circle only when seen from the angle connecting (000) and (111) as illustrated in Fig. 7.49(b).

The ideal demodulation waveforms for the three terminals and a line-to-line voltage are shown in Fig. 7.50. The curve for each terminal is quite

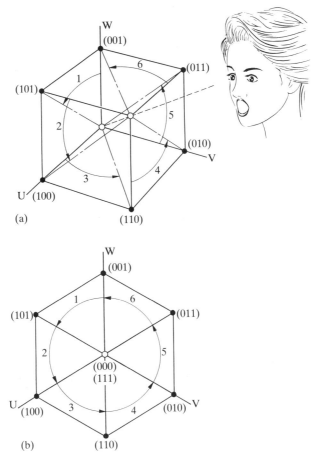

Fig. 7.49. (a) An example of vector locus travelling on the cube surface, and (b) when it is seen along the direction connecting (000) and (111).

different from a sine wave, though the line-to-line waveform is sinusoidal.

Two examples of time sequence for t_0, t_1, and t_2 for each division are given in Fig. 7.51. Example (a) is for the sequence t_o, t_1, and t_2, while in example (b) t_o is divided into two and placed at the beginning and end. It should be noted that the potential at terminal V does not change while the other two are pulsating. Thus, in this method, two of the three phases are always pulse-width modulated, and the two devices belonging to the remaining phase are not. This situation is seen more clearly in Fig. 7.52(a), which is the P W M waveform generated in the former sequence, the number of pulses being 30 per

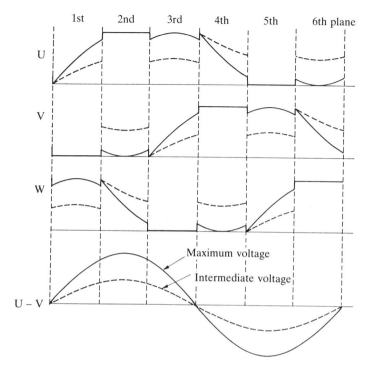

Fig. 7.50. Ideally demodulation curves of each phase and the line-to-line voltage between phases U and V. The solid curves are for the maximum modulation and the broken curves for an intermediate modulation.

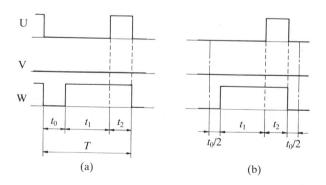

Fig. 7.51. Two examples of time sequence for t_0, t_1, and t_2.

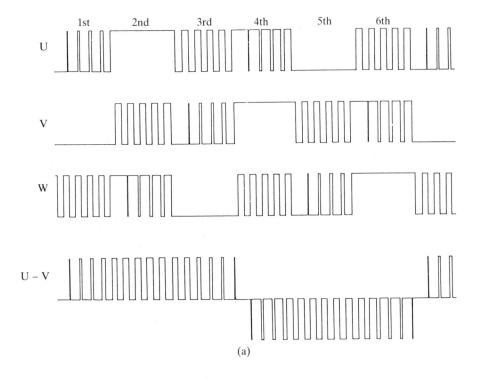

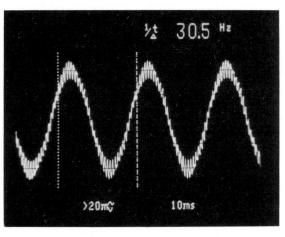

(b)

Fig. 7.52. PWM waveforms generated by a cubic surface method, and current waveforms. (a) Voltage waveforms at each phase terminal; (b) current waveforms in a solid-steel eddy current motor.

Fig. 7.53. Using mechatronic computer KENTAC MARK II for controlling the MECHATRO LAB

cycle. Figure 7.52(b) is the current waveform when an eddy-current motor is driven.

The program for producing such switching signals are not particularly complicated if one uses the idea of six-bit rotation discussed in previous examples.

For effective experiments at this level and more advanced software, the author and his colleagues have developed a special controller named KENTAC MARK II. This controller is shown in the photograph of Fig. 7.53, in which students are running a motor on the MECHATRO LAB.

References

1. Siskind, C.S. (1959). *Electrical machines*, 2nd edn., Chapter 7. McGraw-Hill, New York.
2. Lee, R.H. (1987). Optimal switching to eliminate harmonics in power conversion. *Proceedings of the 14th PCI International Conference, SATECH '87*, pp. 360–72.
3. Taniguchi, K. and Irie, H. (1985). A modulation signal for three-phase sinusoidal PWM inverter. (In Japanese). *Trans. IEE Japan*, **105-B** (No. 10), pp. 76–81.
4. Grant, D.A. and Houldsworth, J.A. (1983). A new high quality PWM AC drives. *IEEE Trans. Ind. Appl.* **IA-19** (Mar./Apr.), pp. 228–34.

8 Brushless direct-current drives

An extension of the inverter is the brushless d.c. drive of a permanent-magnet synchronous motor, in which the switching devices in the inverter portion are controlled by referring to the feedback signal coming from the position sensor. This chapter will start with a plain brushless d.c. motor operated by simple software, and discussions will be extended to drive software, some unique motor structures, and also to the switched-reluctance drive using salient-poled steel rotor instead of a magnet rotor. The brushless d.c. servo-amplifier, or a combination of an inverter and control electronics of a permanent-magnet motor, used for a position control, will also be dealt with.

8.1 Configuration of the brushless direct-current motor system

Details of brushless d.c. motor structures are described in Reference [1]. In Section 1.5. the fundamental principle of using a microprocessor in the brushless d.c. drive of a permanent-magnet motor was explained. We will study this technique more in detail here.

8.1.1 Brushless direct-current drive system

In the inverter drive of a squirrel-cage rotor, the switching signals were applied to the transistor circuit from an external signal generator. If one attempts to drive a permanent-magnet rotor using a normal inverter, one will find that the rotor will not start, unless the initial frequency is extremely low. However, if a position sensor is provided in the motor and switching signals are generated with the correct timing in accordance with the position information, the rotor will start and accelerate like a conventional d.c. motor. This method of driving a permanent-magnet motor is referred to as the brushless d.c. motor. The brushless d.c. motor is supplied with electric current from a d.c. power source and is very similar to the conventional d.c. motor in its torque-versus-speed characteristics, and has the advantage that it is free from brush-wear problems.

As discussed in Chapter 1, a microcomputer can be used for generation of switching signals: and the brushless d.c. motor system is illustrated in Fig. 8.1. The only difference from the inverter system discussed in the previous chapter is that a set of Hall sensors is mounted in the motor, to be used as the position sensor.

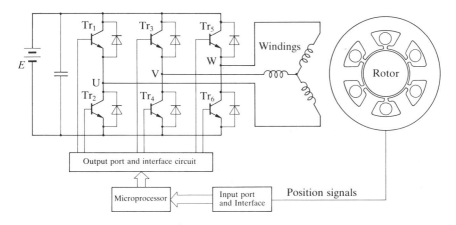

Fig. 8.1. Brushless d.c. drive system using a three-phase bridge inverter, micro-processor, and position sensors.

Although either delta or star connection can be used in the brushless d.c. drive, in this chapter we will use only the star connection. The 180° and 120° conduction methods are the basic schemes for switching sequence of the six-step inverter.

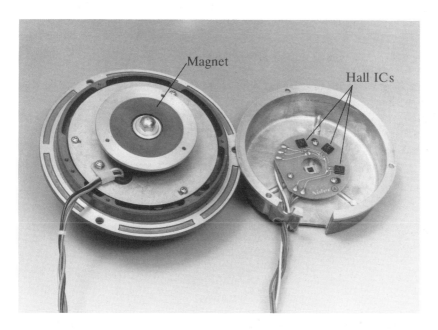

Fig. 8.2. Hall position sensor ICs and magnets in a brushless d.c. motor.

There are brushless d.c. motor systems that do not use any position sensors as such and get the position information from the back e.m.f. appearing in the coils. Reference [2] is recommended to those who are interested in studying its principles. Such types will not be dealt with in this book.

8.1.2 *Hall elements as position sensor*

Among several sensors for detecting the rotor's angular position, the Hall element is the most widely used. In some motors, the rotor magnet is also

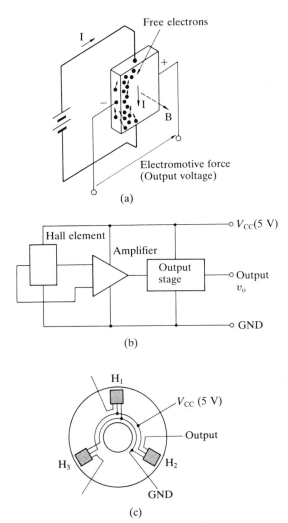

Fig. 8.3. (a) Principle of the Hall element. (b) Block diagram of a Hall IC. (c) Arrangement of three Hall ICs for a three-phase motor.

used as the flux source to expose the Hall elements, while some possess a separate magnet for this purpose, as shown in the photograph of Fig. 8.2.

(1) *Principles of Hall elements.*   Figure 8.3(a) illustrates the principle of the Hall sensor. A typical semiconductor material used for the Hall element pellet is N-doped InSb; it is necessary for a current always to flow in it when it is used as a flux detector. If the pellet is exposed to a magnetic field as depicted, an electromotive force acts on the moving electrons according to Fleming's left-hand rule. Owing to this force, the electrons are directed to the left-hand side, which results in negative polarization on this side and positive polarization on the other side. The electrostatic polarity depends on whether the pellet is passing a North pole or a South pole, and provides the signal to indicate the rotor's position.

(2) *Hall I Cs.*   When Hall sensors are used as the position sensor in a brushless d.c. motor, all the necessary elements are fabricated in an integrated chip in the arrangement shown in Fig. 8.3(b). If the output level is H when the Hall I C is exposed to a North pole, the output level will be L when it is placed in the neighborhood of a South pole. Let us take the case in which three Hall

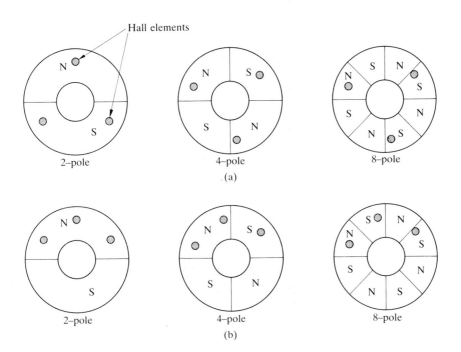

Fig. 8.4. Number of magnetic poles and arrangement of Hall elements: (a) 120° interval; (b) 60° interval.

ICs are used for driving a three-phase motor. In principle, the sensors should be placed 120° apart in a two-pole motor as shown in Fig. 8.3(c). However, as explained later using Fig. 8.13, they can be placed at 60° intervals as shown in Fig. 8.4. It is seen that the 120° and 60° arrangements are effectively the same in a motor for which the number of poles is four or more.

(3) *Interfacing between Hall ICs and the microprocessor.*   Figure 8.5 shows an example of connecting Hall IC output signals to an I/O port. The three output terminals are directly connected to some three bits of an input port, which are pulled by 5 V via resistors.

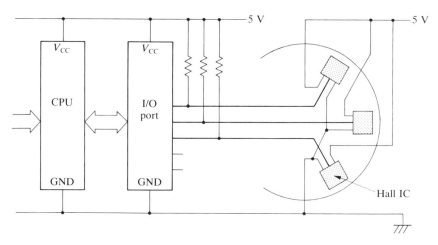

Fig. 8.5. Connection of Hall elements to the output port.

8.2 Six-step 180-degree conduction drives

In studying the application of a microprocessor to drive a brushless d.c. motor, we shall start with a very simple case. Figure 8.6 illustrates a possible relation between the Hall IC positions, coil positions, and drive signals to be fed to the transistors for the CW and CCW directions, respectively. This is clearly the 180° drive of a star-connected motor.

Let us first consider the CW rotation case in Fig. 8.6(a) in which the switching state starts from state (1). The Hall elements are numbered 1, 2, and 3. The upper half of the rotor is the South pole and the lower half is the North pole. In this state, terminals U and W are connected to the positive terminal E and terminal V to the negative, so as to build the current distribution that produces the maximum torque.

In the same figure for state (1) the switching states of the transistors that provide this potential arrangement are also shown. The rotor covers $\pm 30°$ in this state. That is, when the rotor travels 30°, the output state of the Hall I Cs will be state (2), and the rotor will travel another 60° in this new state. The states following this are illustrated in sequences (3), (4), (5), and (6). The relation between the output codes from the Hall elements and the switching state codes is given by Table 8.1(a).

Table 8.1(a)

(a) C W rotation

Hall I C			Switching states					
#3	#2	#1	Tr_6	Tr_5	Tr_4	Tr_3	Tr_2	Tr_1
0	1	0	0	1	1	0	0	1
1	1	0	1	0	1	0	0	1
1	0	0	1	0	0	1	0	1
1	0	1	1	0	0	1	1	0
0	0	1	0	1	0	1	1	0
0	1	1	0	1	1	0	1	0

Table 8.1(b)

(b) C C W rotation

Hall I C			Switching states					
#3	#2	#1	Tr_6	Tr_5	Tr_4	Tr_3	Tr_2	Tr_1
0	1	0	1	0	0	1	1	0
0	1	1	1	0	0	1	0	1
0	0	1	1	0	1	0	0	1
1	0	1	0	1	1	0	0	1
1	0	0	0	1	1	0	1	0
1	1	0	0	1	0	1	1	0

From the relations shown in Fig. 8.6(b), the relation for the C C W rotation is as shown in Table 8.1(b). It should be noted that this correspondence for the C C W movement is produced by inverting each bit of the Hall sensor codes in Table 8.1(a) for the C W case.

8.2.1 *Method of using three bits for driving the power transistors*

In interfacing the inverter and the output port, there are basically two methods for 180° operation. One is to use one bit signal for one phase, so that only three bits are needed, and the other is to use one bit for one power device, so that a total of six bits are used. We shall start with the former method.

(1) *Interfacing.* The I/O port for this drive might be as shown in Fig. 8.7. The A port of an 8255A chip is used as the input port that accepts the Hall I C

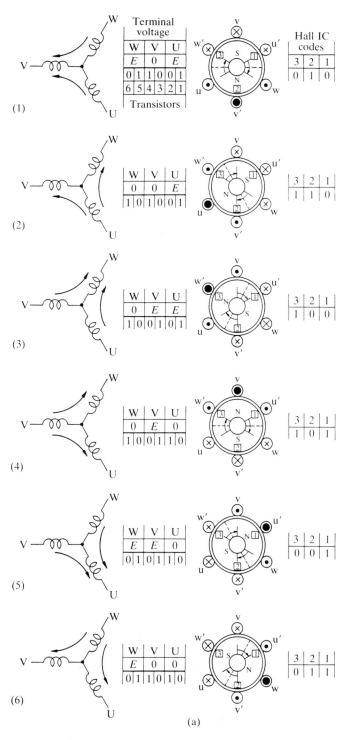

Fig. 8.6. (a) Relation between terminal potentials, drive signals, and Hall IC codes for CW drive of the 180° conductive drive.

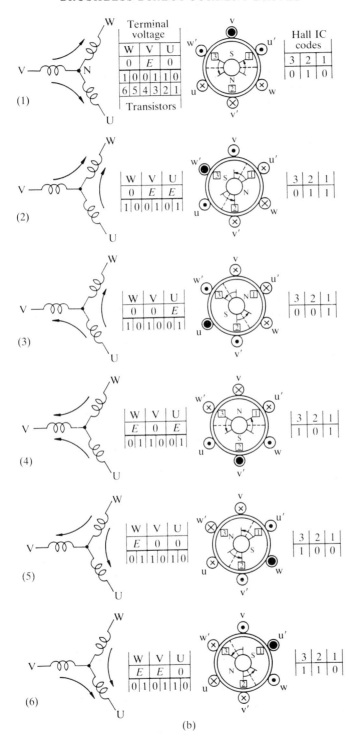

Fig. 8.6 (b) Relation between terminal potentials, drive signals, and Hall IC codes for CCW drive of the 180° conductive drive.

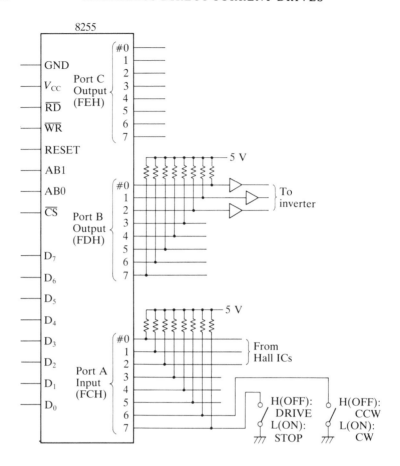

Fig. 8.7. Connection of position-sensor signals to the input port, and connection of output port signals to the inverter.

signals and drive commands. The address of the A port is supposed to have been set to 0FCH. The output port is the B port, whose address is 0FDH, and the lower three bits are connected to the inverter with the circuit configuration for each phase shown in Fig. 8.8. The correspondence between the bit numbers and motor terminals is supposed to be:

#0 bit is for the drive of terminal U;
#1 bit is for the drive of terminal V;
#2 bit is for the drive of terminal W.

In Fig. 8.8, when the H level signal is applied to the input terminal, Tr_1 turns ON, which makes the power transistor Tr_5 turn ON and Tr_4 OFF. When the

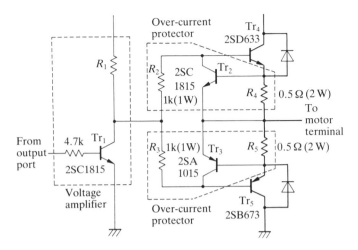

Fig. 8.8. Details of one phase of the inverter.

L level signal is applied to the input terminal, Tr_1 is OFF and as a result Tr_4 is turned ON and Tr_5 OFF.

(2) *Relation between switching states and the output signal for driving a motor.* Let us find the relation between the Hall IC signals and the output port signals to be applied to the inverter input terminals to drive the motor. We recognize that if an output bit is on the H level, the corresponding motor terminal is at GND, and if the output level is the L level, the motor terminal is at potential E. For example, in the state (1) for the CW movement in Fig. 8.6(a), Tr_1 is ON and Tr_2 is OFF; this causes terminal U to be at potential E, and hence the port bit #0 must be at the L level (or state 0). Since Tr_3 is OFF and Tr_4 is ON, terminal V will be at GND, and hence the state of the #1 bit must be H or 1. Using this rule, from the previous two tables, we can derive Table 8.2.

Table 8.2

(a) CW rotation						(b) CCW rotation					
Hall elements			Output signals			Hall elements			Output signals		
#3	#2	#1	W	U	V	#3	#2	#1	W	U	V
0	1	0	0	1	0	0	1	0	1	0	1
1	1	0	1	1	0	1	1	0	0	0	1
1	0	0	1	0	0	1	0	0	0	1	1
1	0	1	1	0	1	1	0	1	0	1	0
0	0	1	0	0	1	0	0	1	1	1	0
0	1	1	0	1	1	0	1	1	1	0	0

From Table 8.2 we can see the following.

(i) For CW movement, the codes from the Hall ICs are identical to the bit state of the output port.
(ii) For CCW movement, the inverse of the Hall IC codes form the output port state.

(3) *Program.*   A simple program is shown in the flowchart of Fig. 8.9. In this program the starting address is labelled START. At this stage, the data fetched to the input port is examined and it is first determined whether the command is DRIVE or STOP. When STOP is instructed, all the three bits are set at the H level and transmitted to the inverter portion to ground the three terminals. Owing to this short-circuiting effect, an effective retarding torque will be produced to decelerate the revolving rotor to an eventual stop.

When #7 bit is on the H level, the interpretation is that the motor is to be run, and the program proceeds to the next step to determine whether CW or CCW is instructed. When CW is instructed, the data in register A, which is the data taken from the input port, is sent out through the output port. After this procedure the program returns to START to repeat the routine. When CCW is instructed, the data in register A is reversed and sent out. After this, the program returns to START. Table 8.3 gives the program listing.

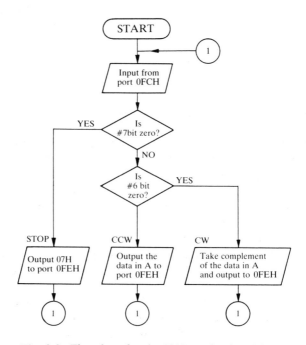

Fig. 8.9. Flowchart for the 180° conduction drive.

Table 8.3. Program listing for driving a brushless d.c. motor using three-bit signals in 180° operation.

```
;**** BRUSHLESS DC MOTOR ****
;       (3-BIT OPERATION)

                        ORG    8100H

00FC            SIGNAL  EQU    0FCH
00FD            POWER   EQU    0FDH

8100   DB FC    START:  IN     A,(SIGNAL); Fetch drive command and
                                        ; Hall sensor code
8102   CB 7F            BIT    7,A       ; See bit #7, and
8104   CA 8117          JP     Z,STOP    ; if zero go to STOP

8107   CB 77    DRIVE:  BIT    6,A       ; See bit #6, and
8109   CA 8111          JP     Z,CCW     ; if zero go to CCW

810C   D3 FD    CW:     OUT    (POWER),A ; Output to drive motor
810E   C3 8100          JP     START     ; Go to START

8111   2F       CCW:    CPL              ; Invert Hall sensor code
8112   D3 FD            OUT    (POWER),A ; Output to drive motor
8114   C3 8100          JP     START     ; Go to START

8117   3E 07    STOP:   LD     A,07H     ; Load A with 00000111B, and
8119   D3 FD            OUT    (POWER),A ; output to brake motor
811B   C3 8100          JP     START     ; Go to START

                        END
```

8.2.2 Using six bits for driving the inverter

Another method of interfacing is the use of six bits for driving the six transistors in the inverter. A typical arrangement of the power portion and interface to the microprocessor is the transistor switching circuit provided in MECHATRO LAB (see Fig. 1.4); the I/O port is illustrated in Fig. 8.10.

For this drive we use a look-up table to store the relation between the Hall sensor codes and switching states given in Table 8.1(a). One can then assemble a program that uses this table. It is arranged such that the lower three bits of each byte in addresses 8000H through 8007H are for the Hall IC output codes, and the memory data are the transistor's switching states. Table 8.4 shows this arrangement.

When CW drive is commanded, the Hall IC codes are received from port F9H. For example, if this code is 011 or 03H, the switching code is 011010 or 1AH and this must be sent out through the output port. Likewise, if 001 is received, 010110 or 16H should be generated.

Table 8.4.   Switching state data stored in the memory area of 8000H through 8007H.

Address	180°	120°(a)	120°(b)
8000H	(00H)	(00H)	09H
8001H	16H	21H	21H
8002H	19H	06H	(00H)
8003H	1AH	24H	24H
8004H	25H	18H	18H
8005H	26H	09H	(00H)
8006H	29H	12H	12H
8007H	(00H)	(00H)	06H

Note: (00H) can be any data.

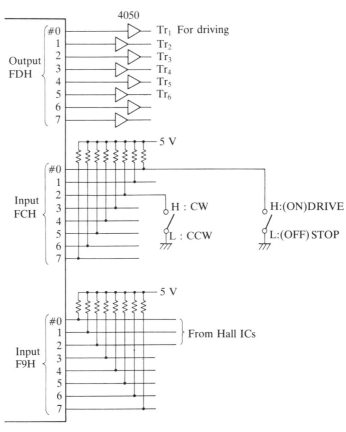

Fig. 8.10. Connecting Hall I Cs output signals and drive command signals to the input port

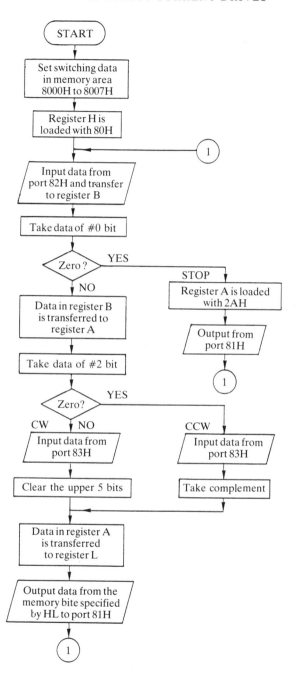

Fig. 8.11. Flowchart for the 120° drive.

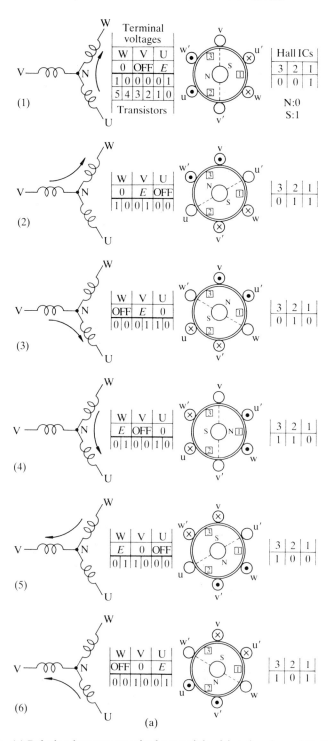

(1)

Terminal voltages		
W	V	U
0	OFF	E

1	0	0	0	0	1
5	4	3	2	1	0

Transistors

Hall ICs		
3	2	1
0	0	1

N:0
S:1

(2)

W	V	U
0	E	OFF

1	0	0	1	0	0

3	2	1
0	1	1

(3)

W	V	U
OFF	E	0

0	0	0	1	1	0

3	2	1
0	1	0

(4)

W	V	U
E	OFF	0

0	1	0	0	1	0

3	2	1
1	1	0

(5)

W	V	U
E	0	OFF

0	1	1	0	0	0

3	2	1
1	0	0

(6)

W	V	U
OFF	0	E

0	0	1	0	0	1

3	2	1
1	0	1

(a)

Fig. 8.12. (a) Relation between terminal potentials, drive signals, and Hall IC codes for a CW drive in the 120° conduction mode.

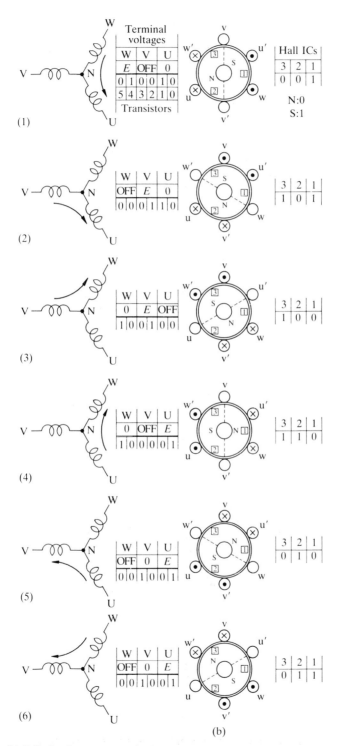

Fig 8.12. (b) Relation between terminal potentials, drive signals, and Hall IC codes for a CCW drive in the 120° conduction drive.

Table 8.5. Program listing for driving a permanent-magnet brushless d.c. motor. The current data stored in the memory area of 8000H through 8007H are for 180-degree operation. By interchanging these data with (00H, 21H, 06H, 24H, 18H, 09H, 12H, and 00H) according to Table 8.4, 120° operation is available.

```
                                ;**** PERMANENT MAGNET BRUSHLESS DC MOTOR ****

                                   ORG    8010H
00F9                        HALL   EQU    0F9H
00FC                        CMD    EQU    0FCH
00FD                        DRV    EQU    0FDH

8010    26 80       HISET:  LD     H,80H
8012    DB FC       START:  IN     A,(CMD)  ;Fetch drive command and
8014    47                  LD     B,A      ;store in B, too
8015    E6 01               AND    1        ;If LSB is zero
8017    CA 8037             JP     Z,STOP   ;go to STOP

801A    78          DRIVE:  LD     A,B      ;Move drive command to A
801B    E6 04               AND    4        ;If bit #2 is zero
801D    CA 802B             JP     Z,CCW    ;go to CCW

8020    DB F9       CW:     IN     A,(HALL) ;Fetch Hall signal code
8022    E6 07               AND    7        ;Clear higher 5 bits
8024    6F                  LD     L,A      ;Move Hall code to L
8025    7E                  LD     A,(HL)   ;Load A with the code
                                            ;stored in (HL)
8026    D3 FD               OUT    (DRV),A  ;Output siwtching code
8028    C3 8012             JP     START    ;Go to START

802B    DB F9       CCW:    IN     A,(HALL) ;Fetch Hall signal code
802D    2F                  CPL             ;Invert code
802E    E6 07               AND    7        ;Clear higer 5 bits
8030    6F                  LD     L,A
8031    7E                  LD     A,(HL)
8032    D3 FD               OUT    (DRV),A
8034    C3 8012             JP     START

8037    3E 2A       STOP:   LD     A,2AH    ;Load A with 00101010B and
8039    D3 FD               OUT    (DRV),A  ;output it to brake motor
803B    C3 8012             JP     START    ;Go to START

                                   ORG    8000H
8000    00 16 19 1A DATA:   DB     00H,16H,19H,1AH ;Switching codes
8004    25 26 29 00         DB     25H,26H,29H,00H ;Switching codes

                                   END
```

When CCW drive is commanded, the Hall IC codes are inverted after being received in the CPU, and the switching codes are generated with reference to this inverted code. For example, if the input code is 001, it must be converted to 110, and the corresponding switching code is 101001 or 29H.

Figure 8.11 is the flowchart and Table 8.5 is its program listing. In this program, four registers are used for the following purposes:

A: receiving and sending data from/to the I/O port;
B: temporary memory for the drive command and the Hall IC output codes;
H: 80H is always stored in this register to specify the upper byte of the look-up table address;
L: the lower byte of the look-up table address is stored in this register.

8.3 Six-step 120-degree conduction drives

Most brushless motors used in office automation equipment employ the 120° conduction type for the star-connected windings. Figure 8.12 illustrates the relation between the positions of the Hall elements, current distribution, the rotor position, and the switching codes. Note that the positions of the Hall sensors are different from the case of Fig. 8.6 for the 180° conduction scheme. The relation between the Hall sensor codes and switching codes for this case is given in column (a) in Table 8.4. It is evident that the same

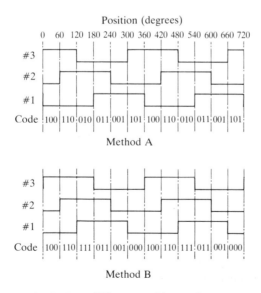

Fig. 8.13. Two different position code systems.

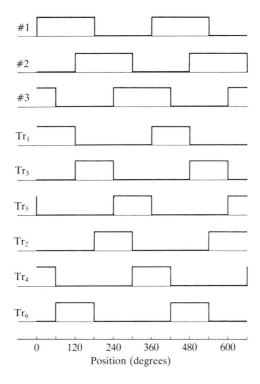

Fig. 8.14. Relationship between the Hall IC codes and transistor states.

program for the 180° drive is usable only by changing the data in the look-up table.

In practice, there are two ways of encoding the Hall sensor output states, as illustrated in Fig. 8.13. Method A is the encoding method so far discussed. When the output codes are generated according to method B, the data in the look-up table should be changed as shown in column (b).

Figure 8.14 illustrates the time sequence of the Hall sensor codes in method A and the switching signals to drive the transistors. When a hardware logic circuit is to be employed to generate such relations, the example given in Fig. 8.15 can be used.

8.4 Unique brushless direct-current motor and its drives

In Section 1.3.1 it was stated that the number (q) of slots per phase per pole, which is given by the number of slots/(number of poles multiplied by number of phases), must be an integer and that it is desirable that it be greater than

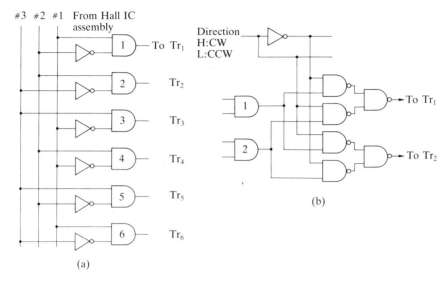

(a)

(b)

Fig. 8.15. Hardware logic for generating 120° drive signals from Hall IC output codes. (a) Logic for unidirectional drive; (b) additional logic for reversing. Two more identical circuits are needed for the pairs of Tr_3/Tr_4 and Tr_5/Tr_6.

one. In office automation equipment, however, motors to which this rule does not apply are used in large quantity.

8.4.1 *Axial flux slotless motor*

Let us start with the motor shown in the photograph of Fig. 8.16, which is the spindle motor used in a floppy-disc drive. Six coils are arranged at 60° intervals on a plane and the rotor magnet is magnetized in eight poles as illustrated in Fig. 8.17(a). A simpler model is the 4-pole 3-coil motor shown in (b). This axial-flux machine can be converted into a radial flux machine as shown in Fig. 8.18. Three Hall elements are placed in the centre of coils.

Figure 8.19 shows the relation between rotor position and current distribution as the motor travels in a CW direction. It is obvious that the program listed in Table 8.5 can be used to run this motor. A drawback of this construction is that torques are not produced at symmetrical positions with respect to the rotor shaft. In the 8-pole 6-coil structure there are no problems.

8.4.2 *Motor with fewer than one slot per phase per pole*

The slotted-core version of the radial-flux 8-pole machine is available using the stator B in the MECHATRO LAB presented in Chapter 1. The wind-

Fig. 8.16. Disassembled flat motor of the axial-flux type having six coils.

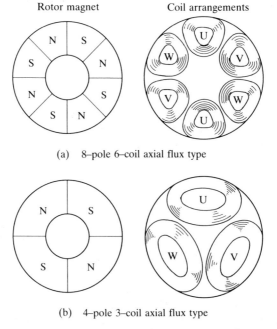

Fig. 8.17. Eight-pole six-coil, and four-pole three-coil arrangements

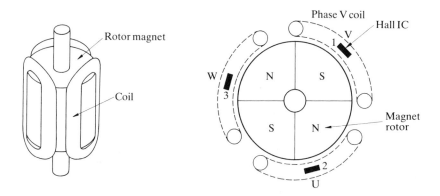

Fig. 8.18. Four-pole three-coil arrangement for the radial flux machine.

ing is much simpler than with the 24-slot motor discussed in Section 1.3.1. The number q with this simple motor is $1/4$. Although this structure is less efficient than an orthodox motor, it is employed in great numbers in information-handling machines such as laser printers.

Another irregular but useful construction is the 4-pole 6-slot motor. Figure 8.20 shows how this motor can be operated with an inverter.

8.4.3 *Toroidal-coil brushless direct-current motor*

The toroidal-coil brushless motor is also an interesting device. The basic construction is shown in Fig. 8.21. See Fig. 8.22 for the relation between the rotor positions and the current distribution for a star-connected motor driven in the 120° conduction mode. Suitable applications of this type motor are found in high-speed use, because at high speeds the back e.m.f. is so high that the number of turns must be few, and a single-layer coil structure is suitable that is easy to manufacture.

8.4.4 *Brushless direct-current drive of a salient-poled solid-steel motor*

By replacing the permanent-magnet rotor with a salient-poled rotor of either the squirrel-cage or solid-steel type, we can construct a unique brushless d.c. motor. This can be regarded as the closed-loop drive of a stepping motor[3] or as a switched reluctance drive proposed by P.J. Lawrenson[4] as an energy-saving drive. A basic difference is seen in the means of reversing and braking. The suitable positions of the Hall sensors with respect to the winding positions are also different from the normal permanent-magnet brushless drive.

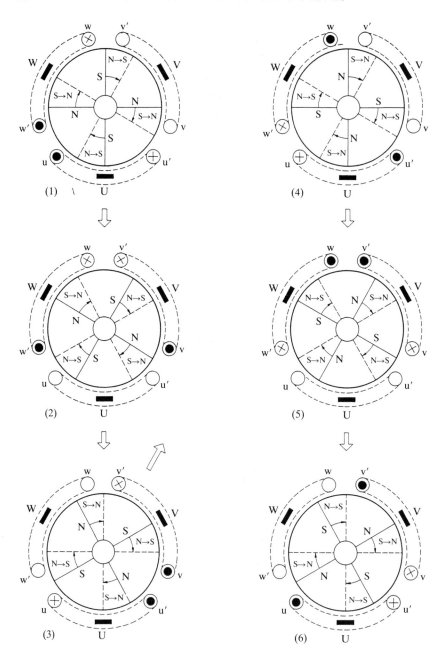

Fig. 8.19. How the three-coil four-pole motor rotates.

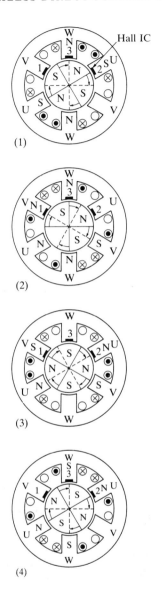

Fig. 8.20. Rotation of a motor having a fractional index of slots/phase/pole.

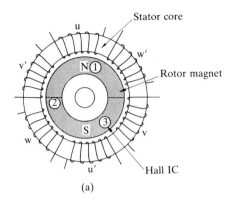

(a)

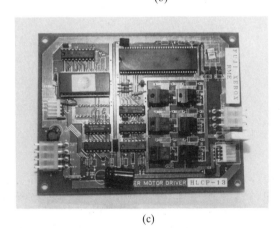

(b)

(c)

Fig. 8.21. Brushless d.c. motor with a ring core and toroidal coils. (a) Schematic diagram; (b) a motor designed for a polygon mirror drive; (c) the control/drive circuit. (By courtesy of Manufacturing Engineering Laboratory, Fuji Xerox Co., Ltd.)

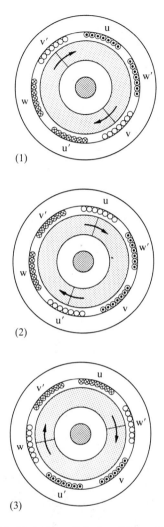

Fig. 8.22. Rotation of the toroidal-coil motor.

Figure 8.23 shows how to construct an experimental set-up for brushless d.c. drive of a solid-steel reluctance motor. The six-toothed stator with concentrated windings is used here, too. The relations between the rotor positions and the current distribution in the coils are illustrated in Fig. 8.24 for (a) CW rotation, (b) braking, and (c) CCW rotation. It is seen that this drive corresponds to the two-phase-on drive of the stepping motor. It is also obvious that the short-circuiting brake is not effective for a rotor that has no permanent magnets. Consequently, in decelerating the rotor, switching

Fig. 8.23. Constructing a brushless d.c. motor using a solid-steel salient-poled rotor and a six-toothed stator with concentrated windings.

signals are generated so that the rotor has a tendency to stall at the current position in the example of Fig. 8.24(b).

Figure 8.25 is the flowchart of the program for generating the switching signals in reference to the rotor position information. Table 8.6 is the program listing for this drive.

8.5 Brushless servo-amplifiers

The power-handling portion of a positioning system that uses servomotors is referred to as a servo-amplifier. A servo-amplifier designed for driving a brushless d.c. motor is a type of inverter. A typical construction of a brushless servomotor is the permanent-magnet synchronous motor shown in Fig. 8.26. As with an induction motor, the armature windings, which are installed in the stator core, are of three-phase delta or star connection, but the rotor has permanent magnets. Unlike the case with brushless motors employed in office-automation equipment, the number of slots per phase per pole is usually 2 or more. Some typical rotor constructions are shown in Fig. 8.27.

A fundamental type of servo-amplifier for a brushless d.c. motor is shown

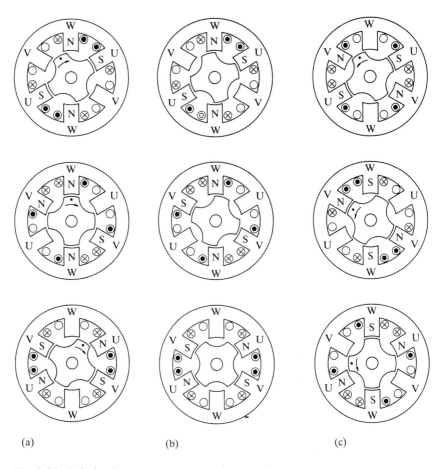

(a)   (b)   (c)

Fig. 8.24. Relation between the rotor positions and current distribution in the coils for (a) CW rotation, (b) braking, and (c) CCW rotation.

Fig. 8.25. Flowchart for generating switching signals for brushless d.c. drive of a reluctance motor.

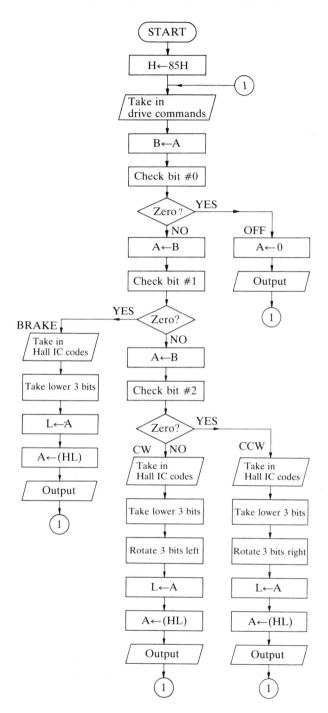

Table 8.6. Program listing for the brushless reluctance motor drive.

```
;  ***** BRUSHLESS RELUCTANCE MOTOR *****

                       ORG     8508H

              HALL     EQU     0F9H        ;Output port for HALL IC code
              COMD     EQU     0FCH        ;Input port for drive command
              DRIV     EQU     0FDH        ;Output port for driving motor

8508  26 85   HISET:   LD      H,85H       ;Set higher byte address for
                                           ;drive command
850A  DB FC   START:   IN      A,(COMD)    ;Fetch drive command
850C  47               LD      B,A         ;and load it in B
850D  E6 01            AND     1           ;See LSB (bit #0)
850F  CA 8548          JP      Z,OFF       ;and if zero, then go to OFF

8512  78      STOP:    LD      A,B
8513  E6 02            AND     2           ;See bit #1 and
8515  CA 854F          JP      Z,BRAKE     ;if zero, then go to BRAKE

8518  78      DIRECT:  LD      A,B
8519  E6 04            AND     4           ;See bit #2
851B  CA 8533          JP      Z,CCW       ;and if zero, then go to CCW

851E  DB F9   CW:      IN      A,(HALL)    ;Fetch Hall IC code
8520  E6 07            AND     7           ;Clear higher 5 bits
8522  4F               LD      C,A         ;Rotate 3 bits left
8523  E6 04            AND     4
8525  0F               RRCA
8526  0F               RRCA
8527  0F               RRCA
8528  81               ADD     A,C
8529  07               RLCA
852A  E6 07            AND     7           ;Clear higher 5 bits
852C  6F               LD      L,A         ;Road the result in L
852D  7E               LD      A,(HL)      ;Load A with switching code
852E  D3 FD            OUT     (DRIV),A    ;Output
8530  C3 850A          JP      START       ;Go to START

8533  DB F9   CCW:     IN      A,(HALL)
8535  E6 07            AND     7
8537  4F               LD      C,A
8538  E6 01            AND     1
853A  07               RLCA
853B  07               RLCA
853C  07               RLCA
853D  81               ADD     A,C
853E  0F               RRCA
853F  E6 07            AND     7
8541  6F               LD      L,A
8542  7E               LD      A,(HL)
8543  D3 FD            OUT     (DRIV),A
8545  C3 850A          JP      START
```

Table 8.6. *continued*

```
8548    3E 00     OFF:     LD     A,0        ;Clear A
854A    D3 FD              OUT    (DRIV),A   ;Open all transistors
854C    C3 850A            JP     START      ;Go to START

854F    DB F9     BRAKE:   IN     A,(HALL)   ;Fetch Hall IC code
8551    E6 07              AND    7          ;Clear higher 5 bits
8553    6F                 LD     L,A        ;Load it in L
8554    7E                 LD     A,(HL)     ;Load A with switching code
8555    D3 FD              OUT    (DRIV),A   ;Output
8557    C3 850A            JP     START      ;Go to START

                  ;*****  SWITCHING CODES *****

                           ORG    8500H

8500    00 21 06 24  DATA:  DB     0,21H,06H,24H,18H,09H,12H,0
8504    18 09 12 00
                           END
```

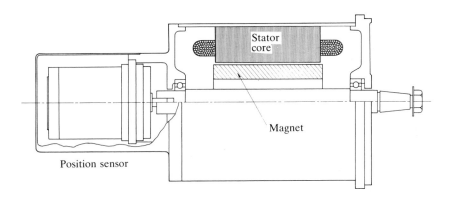

Fig. 8.26. Basic construction of a brushless d.c. servomotor.

in Fig. 8.28. An important difference from a plain inverter employed in the speed adjustment of an induction motor is that the switching signals are produced with reference to the rotor position. Another difference is that this amplifier uses two d.c. power sources and the windings are star-connected, the neutral point being grounded. Hence, the power circuit for each phase is independent of the other two.

Figure 8.29 shows the similarities in and differences between the servo-amplifier for a conventional d.c. motor and that for a brushless d.c. motor. Figure 8.29(a) is the d.c. amplifier supplied by two d.c. sources and (b) shows

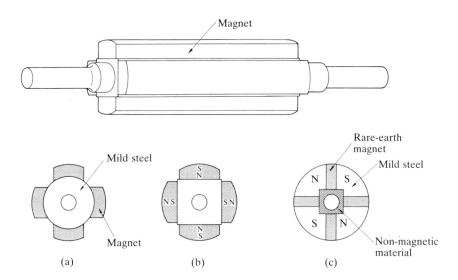

Fig. 8.27. Typical construction of rotors used in brushless d.c. servomotors.

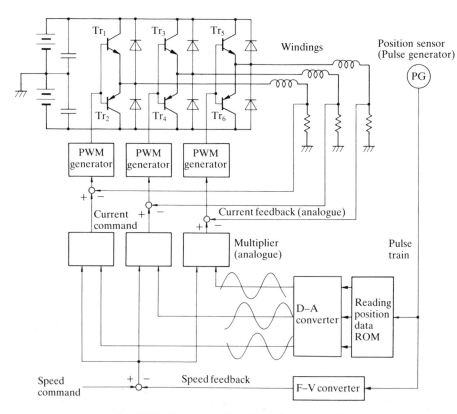

Fig. 8.28. Elementary brushless d.c. servo drive.

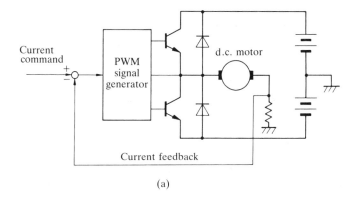

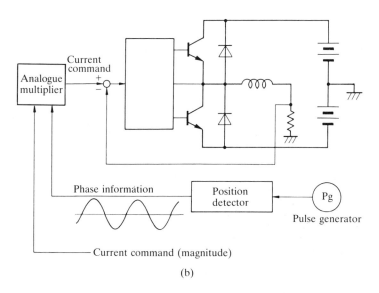

Fig. 8.29. Comparison of servo-amplifiers for (a) a conventional d.c. motor and (b) a brushless d.c. servomotor.

one phase of the brushless d.c. servo-amplifier. They are basically very similar, but the difference is seen in the input signals. In circuit (a), the input is the current command, which may be positive or negative. Circuit (b) has two input terminals; one is for the current amplitude and the other the sine wave signal $\sin \theta$, where θ is the phase related to the rotor position data. The amplitude multiplied by $\sin \theta$, which can be computed by analogue or digital means, is used as the actual current command. For the other two phases, the same amplitude signal is used, and the phase signals are $\sin(\theta + 2\pi/3)$ and $\sin(\theta - 2\pi/3)$.

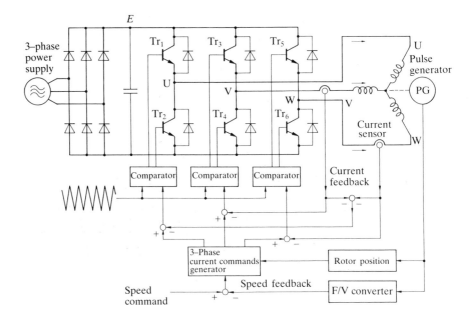

Fig. 8.30. Servo-amplifier that does not use the neutral point.

The data θ should be determined so that the resultant magnetic field produced by the three-phase windings is oriented at right angles to the rotor's magnetic field. If the arrangement is such as to make the angle positive ($+90°$) for a positive current command signal, then the angle will be $-90°$ when the amplitude command is negative.

In practical applications, where d.c. power is to be provided from a commercial three-phase power supply, the circuit configuration requiring two d.c. supplies (in as Fig. 8.28) can be expensive. For this reason, as for an ordinary inverter used for speed control of an induction machine, the scheme shown in Fig. 8.30 is widely employed in brushless d.c. servo-amplifiers. Here the windings are connected in the delta or star scheme. Since in this case the current I_3 flowing in the third phase automatically equals $-(I_1 + I_2)$, only the currents in Phases 1 and 2 need to be detected.

The current amplitude instruction is given as a supervisory command in the servo-system, while the phase instruction for each winding is produced from a look-up table stored in a ROM IC. Here the memory addresses are the rotor positions θ, and their data are $\sin \theta$, $\sin(\theta + 2\pi/3)$, and $\sin(\theta - 2\pi/3)$.

As is obvious, another considerable difference from an ordinary inverter used for speed adjustment of an induction motor is that in a brushless d.c. servo-amplifier the current is controlled, while in the inverter the voltage is controlled.

Fig. 8.31. Brushless d.c. servo-amplifiers and motors. (By courtesy of Sanyo Denki Co. Ltd.)

The photograph of Fig. 8.31 shows a series of brushless d.c. motors and servo-amplifiers using bipolar transistors.

References

1. Kenjo, T. and Nagamori, S. (1985). *Permanent-magnet and brushless DC motors*. Oxford University Press, Oxford.
2. Endo, T., Tajima, F., Iizuka, K., and Uzuhashi H. (1985). Brushless Motor without a shaft-mounted position sensor. *Trans. IEEE Japan* **105** (1/2), 1–7.
3. Kenjo, T. (1985). *Stepping motors and their microprocessor controls*, Chapter 7. Oxford University Press, Oxford.
4. Lawrenson, P. (1985). Switched reluctance drive—a fast growing technology. *Electric Drives & Controls* (April/May), 18–23.

9 General discussion on converters

In the preceding chapters we studied the details of several types of power converters and motor-drive circuitry such as a.c.-to-d.c. converters, d.c.-to-d.c. converters, and the d.c.-to-a.c. converter known as the inverter. Through these studies it was seen that the H-bridge circuit and its variations are of great use in power electronics. But we have not yet surveyed all the functions of this fundamental circuit. First, it will be demonstrated that the H-bridge can also be used as an a.c.-to-a.c. converter known as cyclo-converter. Discussions will be extended to review the single-phase inverter. The second consideration will be the study of a generalized d.c. converter to elaborate the interrelation between step-up and step-down converters and between the d.c. converter and the single-phase inverter.

9.1 Possibilities of the bridge circuit

In the discussions of the preceding chapters we observed that the H-bridge circuit is simple and basic to power electronics. However, we have not yet studied all the fundamental functions of this circuit. Let us examine the possibilities of the bridge circuit more in detail.

9.1.1 Alternating-current to direct-current converter

We start with the circuit of Fig. 9.1(a), in which there are two boxes A and B: one represents the load and the other an a.c. or d.c. power supply. The supply and the load are linked via two double-throw switches that can each have three states: upper ON, down ON, and open. As shown in Fig. 9.1(b), the basic function of each of these switches can be provided via a combination of two single-throw switches that each have two states: ON and OFF. A mechanical switch of this type would be replaced by some type of solid-state switch in power electronics. In Fig. 9.1(c), the bridge is composed of four transistors plus four diodes.

A simple but practical form of this converter is an a.c.-to-d.c. converter, as shown in Fig. 9.2. Here, box B is the a.c. power source and box A is the load, and all the transistors are always OFF. In this circuit, it is impossible for box A to be the a.c. power source and box B the load, because diodes can provide short-circuit paths.

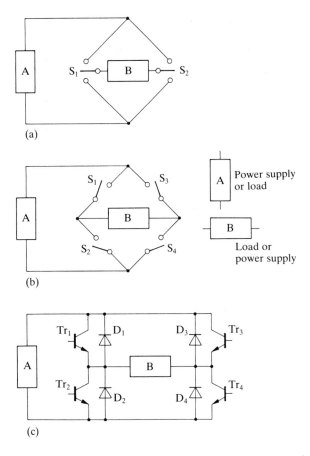

(a)

(b)

(c)

Fig. 9.1. Basic bridge converter circuits. (a) Two boxes, one representing the power supply and the other the load, are connected via two double-throw switches. (b) The two double-throw switches are replaced by four single-throw switches. (c) Each single-throw switch is replaced by a combination a transistor and a diode.

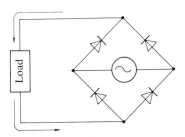

Fig. 9.2. An a.c.-to-d.c. converter is created if all transistors are off or are eliminated from the circuit of Fig. 9.1(c).

9.1.2 *Principle of the cyclo-converter*

A more general circuit is shown in Fig. 9.3(a), in which each mechanical switch has been replaced by two transistors and two diodes. The input or primary side is a single-phase a.c. supply. Figures (b) and (c) show two ways of using these transistors, depending on the supply polarity. In (b) when the

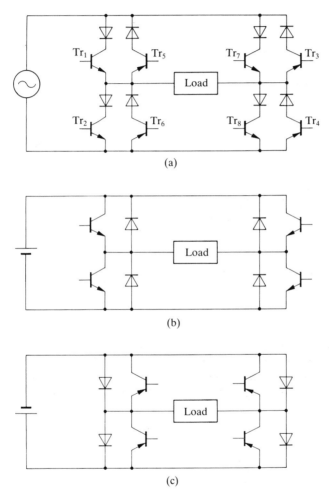

Fig. 9.3. Fundamental cyclo-converter circuit. (a) The switching element is a back-to-back combination of two series circuits of a transistor and a diode, where each transistor is regarded as a unidirectional switch. (b) When the supply polarity is positive, Tr_5 to Tr_8 are closed and the rest of the transistors are used for controlling the load current/voltage. (c) When the supply is negative, Tr_1 to Tr_4 are closed and Tr_5 to Tr_8 are used for controlling the load current/voltage.

power supply is positive, Tr_5 to Tr_8 are all closed while Tr_1 to Tr_4 are used for controlling the current or voltage polarities of the load. In (c) when the supply is negative, Tr_1 to Tr_4 are all closed while the other transistors are for control use.

It should be noted that in this circuit the power supply can either be d.c. or a.c., and also that the load voltage can be d.c. or a.c. When the supply is a.c. of a certain frequency and the load voltage is to be a.c. of a different frequency, this bridge may be defined as the fundamental cyclo-converter.

Figure 9.4 shows several examples of the output waveforms available from this fundamental scheme. It is seen that the output frequency can be either

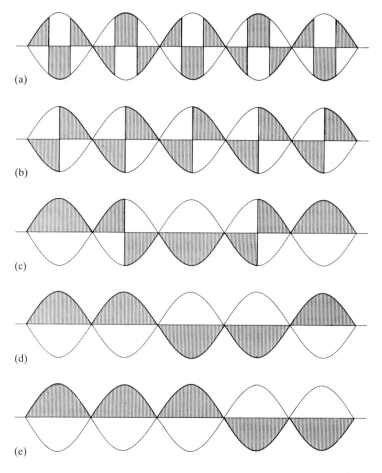

Fig. 9.4. Output waveforms of a cyclo-converter: (a) and (b) are for output frequencies higher than the input; (c) to (e) are for output frequencies lower than the input.

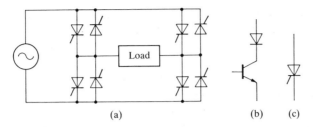

Fig. 9.5. (a) Single-phase input cyclo-converter using thyristors; a combination of a diode and transistor (b) is here replaced by a thyristor (c).

lower or higher than that of the input, although the waveforms contain large proportions of harmonics. If the output frequency is to be lower than the input, thyristors can be used instead of the combination of a transistor and a diode, as shown in Fig. 9.5.

9.1.3 Three-phase-input cyclo-converter

In practice, a single-phase input is not suitable for the cyclo-converter because of high-order harmonics involved in the load voltage. The three-phase input type shown in Fig. 9.6 is a practical scheme. Two sets of three-phase full-bridge converters are used and, hence, this configuration can be termed a dual converter. One set is used for providing positive current, and the other for negative current, similarly to the four-quadrant converter discussed in Section 3.5.

Typical waveforms of the output voltage are shown in Fig. 9.7 in reference to the load current. The current is seen to be leading voltage in phase. In the

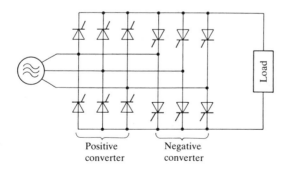

Fig. 9.6. Three-to-single-phase cyclo-converter. Two sets of the full-bridge converters are used; one is for supplying positive current and the other for negative current.

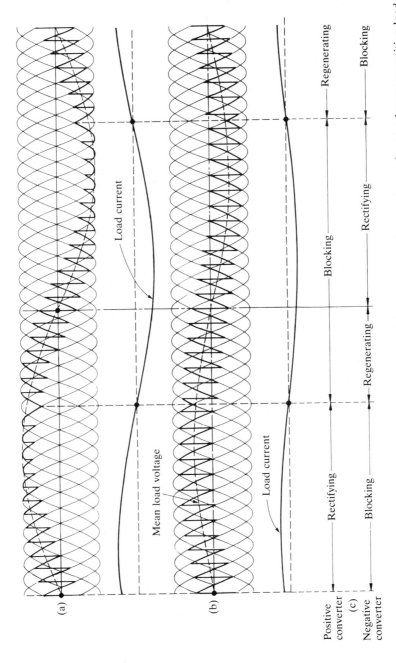

Fig. 9.7. Typical waveforms output by the three-phase input cyclo-converter, and current waveforms when a capacitive load is operated. (a) Trace of the voltage profile when the maximum output voltage is provided. (b) The case of half-maximum voltage. (c) Functions of the positive and negative converters.

region where the polarities of the voltage and current are opposite, electric power is fed back to the power supply, as often occurs in a current-source inverter, which will be discussed in the next section.

When three-phase output is required, three sets of such converters must be used. Accordingly, the three-phase cyclo-converter can be very expensive and consequently inverters are more suitable in low- and medium-power applications.

9.2 Single-phase inverters

A combination of two bridges can constitute a single-phase inverter. Depending on the type of power switches and auxiliary passive elements used, inverters can be divided into two categories.

9.2.1 *Voltage-source inverter*

In the circuit shown in Fig. 9.8, bridge (1) is a diode bridge working as an a.c.-to-d.c. converter, and bridge (2) is the d.c.-to-a.c. converter using transistors and diodes. A capacitor is placed between these two bridges to absorb pulsation voltages after rectification of the input power, and also to keep the d.c. voltage at a constant level when commutation takes place. This is a voltage-source inverter. Figure 9.9 illustrates the current paths during commutation. When the load causes current to feed back towards bridge (1) owing to the inductive load just after Tr_1 and Tr_4 are turned off, the capacitor absorbs the current. It should be noted that the current is not allowed to return to the power supply, flowing from the cathode to the anode through the diodes used in the converter (bridge (1)).

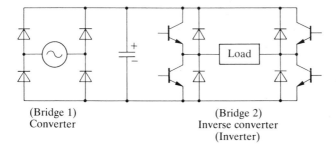

(Bridge 1)   (Bridge 2)
Converter   Inverse converter
                       (Inverter)

Fig. 9.8. Voltage-source inverter as a combination of two bridge circuits. The converter portion uses only diodes and the inverse-converter portion employs transistors and diodes.

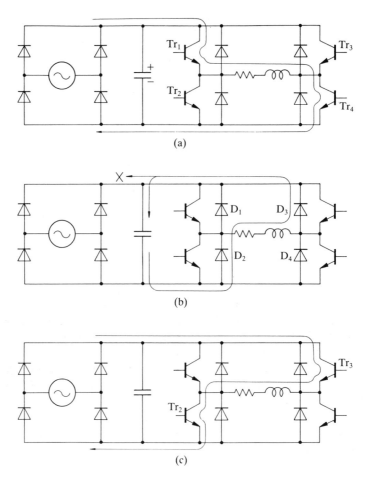

Fig. 9.9. Current paths during commutation in the voltage-source inverter. (a) Tr_1 and Tr_4 are carrying current. (b) Tr_1 and Tr_4 are turned OFF, and Tr_2 and Tr_3 turned ON, the transient current is fed back to the supply side, but it is stored in the capacitor placed across the d.c. lines. (c) When the load current reverses, Tr_2 and Tr_3 carry the current.

9.2.2 Current-source inverter

Another possible inverter comprises a combination of two thyristor bridges, as shown in Fig. 9.10. In this scheme a reactor is placed between the two stages to make the system work as the current-source inverter. The reactor causes the current to flow from stage (1) towards stage (2) during commutation. In the inverse-converter portion (stage (2)), no flyback diodes, but

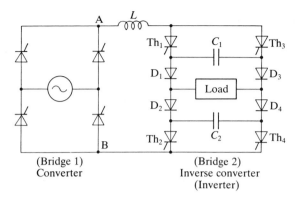

Fig. 9.10. Current-source inverter is set up by using thyristors in both bridge circuits. The inverse-converter portion uses no flyback diodes, but series diodes and capacitors are placed across the cathodes of the thyristors.

capacitors are used for turning off thyristors and also for providing current paths for the transient currents lasting for a short period after turning ON and OFF takes place. The functions of the four diodes placed in series are explained Fig. 9.11 along with the functions of the capacitors.

It should be noted that the voltage polarity at the d.c. stage (between A and B in the figure), which is sandwiched by two a.c. stages, can be reversed by a.c. adjustment of the firing angle at the converter portion. When the voltage is positive, power is transmitted from left to right, while it is transmitted from right to left when the polarity is reversed, as shown in Fig. 9.12.

In practice, the current-source inverter is used as a large three-phase inverter, as explained in Section 7.4.2.

9.3 Generalized direct-current-to-direct-current converter

The circuit in Fig. 9.13 is a generalized d.c.-to-d.c. converter. Through a study of this circuit and its extention to an H-bridge, we can deepen our knowledge of the functions of the bridge circuit.

9.3.1 Interrelation of step-up and step-down converters

The generalized d.c. converter of Fig. 9.13 can be operated with two batteries: voltage E_1 is higher than E_2. Electrical power is transmitted either from the higher-voltage side (E_1) to the lower voltage side (E_2), or vice versa. All the circuit elements are contained in the square box indicated by dotted lines, and these idealized elements do not dissipate power into heat loss.

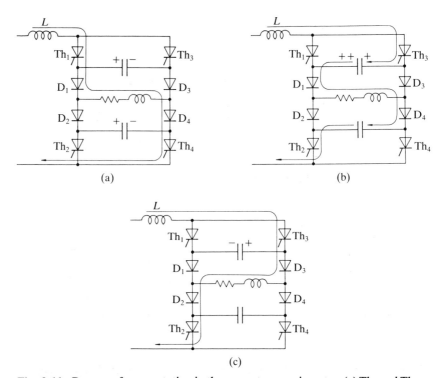

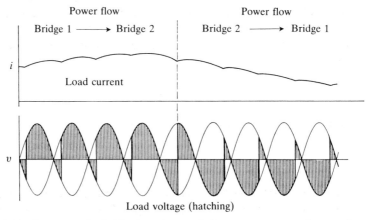

Fig. 9.11. Process of commutation in the current-source inverter. (a) Th_1 and Th_4 are carrying a current. (b) When Th_2 and Th_3 turn ON, Th_1 and Th_4 turn OFF automatically owing to a reverse bias voltage applied via charged capacitors. When the load is inductive, the transient current will flow through the two capacitors and D_1 and D_4 as shown. (c) When the capacitor potential and the load potential become the same, D_2 and D_3 turn ON, and when the load current reverses, D_1 and D_4 will be turned OFF.

Fig. 9.12. Power can be fed back to the supply in a current-source inverter by adjustment of firing angle at the converter portion.

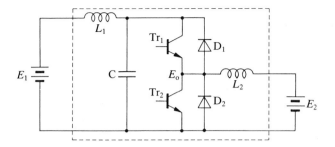

Fig. 9.13. Generalized d.c.-to-d.c. converter.

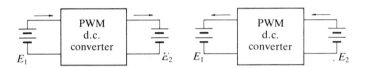

Fig. 9.14. Black box representation of the d.c.-to-d.c. converter.

If we regard the square as a black box, the converter of Fig. 9.13 can be represented by Fig. 9.14. There are two possible cases of power flow in this circuit:

(a) E_1 supplies electric power and E_2 is charged;
(b) E_2 supplies electric power and E_1 is charged.

We shall study these cases when either Tr_1 or Tr_2 is in the ON state and the other is in the OFF. When Tr_1 is closed and Tr_2 is open, potential E_0 at the point in Fig. 9.13 at which both transistors are connected equals E_1, while it is zero when Tr_1 is open and Tr_2 is closed. It is assumed that the forward drop in a transistor is negligibly small. Hence, the the average potential $\langle E_0 \rangle$ is expressed by

$$\langle E_0 \rangle = \frac{E_1 T_{on}}{T_{on} + T_{off}} \tag{9.1}$$

which must be equal to the secondary potential E_2:

$$E_2 = \frac{E_1 T_{on}}{T_{on} + T_{off}}. \tag{9.2}$$

It should be noted that this relation holds independently of the current direction. When electric power is flowing to the right, E_1 is supplying power by stepping down its potential; when power is flowing to the left, E_2 is supplying electric power by stepping up its potential. In the former case, the circuit is

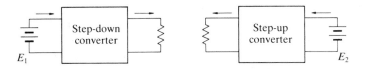

Fig. 9.15. Power flow in the step-up and step-down converters.

acting as a step-down converter and the battery E_2 may be replaced by a normal load. In the latter, the circuit behaves as a step-up converter and the battery E_1 may be replaced by a normal load. These conditions are illustrated in Fig. 9.15.

We shall now examine the roles of the switches for each case shown in Fig. 9.13. First, we take the case in which power flows to the right through reactor L_2. Obviously, when Tr_1 is closed, the current will flow through it. When Tr_1 is open and Tr_2 is closed, however, the secondary current flows not through Tr_2 but through diode D_2. Consequently, switch Tr_2 and diode D_1 are not needed here. We may therefore conclude that the circuit should be simplified as shown in Fig. 9.16(a). Reactor L_2 and capacitor C_2 are needed for providing a smooth d.c. output current. The reactor and capacitor in the primary portion are necessary when the input current is required to be a d.c. current

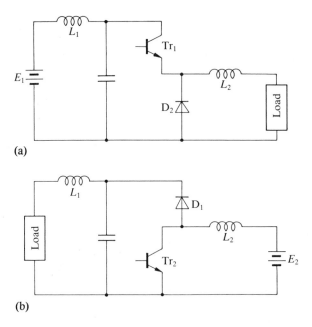

Fig. 9.16. Configurations of (a) the step-up converter and (b) step-down converter.

with little fluctuation. In many cases, however, the reactor L_1 is eliminated, but the capacitor is needed to stabilize the primary side potential.

Next we examine the other case, in which the power source E_2 supplies electric power. In this case, a current is always flowing through reactor L_2 to the left. Therefore, when Tr_1 is closed and Tr_2 is open, the current must flow through diode D_1, not through switch Tr_1, because a transistor can only be used as a unidirectional switch. When Tr_2 is closed and Tr_1 is open, the current flows through Tr_2. Thus, it is seen that Tr_1 and diode D_2 are not needed here, and hence the circuit may be simplified as shown in Fig. 9.16(b). It is seen that in this circuit, reactor L_1 is not always necessary.

9.3.2 Driving a motor as the load

Refer to Fig. 9.17, in which a separately excited d.c. motor is driven using a d.c. converter circuit. In the normal mode, as we have seen, Tr_2 and D_1 are not necessary. However, when the motor is accelerated by some means, the d.c. machine will act as a generator, and current will flow in reactor L_2 from right to left, and transistor Tr_2 and diode D_1 are effectively used to make the circuit work as a step-up converter.

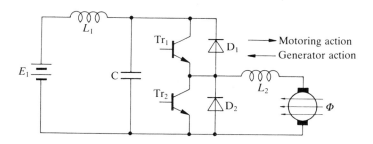

Fig. 9.17. Direct-current motor driven by a d.c. converter.

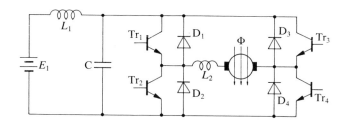

Fig. 9.18. Bridge d.c. converter for driving a separately excited d.c. motor in the reversible mode.

When the motor is to be operated in the reversible mode, it must be driven in the full-bridge scheme as shown in Fig. 9.18. Now, for the forward drive, Tr_4 is closed and Tr_3 is open, and Tr_1 and Tr_2 are operated in the PWM mode, with diode D_2 working as a free-wheeling diode. In this state, if the machine runs in the generator mode, the current will be fed back to the power supply though D_1 and D_4. Similarly, when Tr_2 is always closed and Tr_1 open, and Tr_3 and Tr_4 are used in the PWM mode, the motor runs in the reverse direction. In this state, diode D_4 may be thought of as a free-wheeling diode.

9.3.3 *Four-quadrant operation of a direct-current machine*

As seen above, an H-bridge can operate a d.c. motor in either rotational direction, and also in either the motoring or the generator mode. This capability is often referred to as the four-quadrant operation, and is closely related to the concept as defined in Section 3.5. As illustrated in Fig. 9.19, the vertical axis is taken as the load voltage and the horizontal one as the current.

It should be noted that the current is proportional to the torque in a d.c. motor, and that a major portion of the applied potential constitutes the back e.m.f., which is proportional to the rotational speed. Here, quadrant I is defined as the state in which a positive average voltage drives the machine with a positive current in the CW direction. Hence, if both the voltage and the current reverse, the machine will work as a motor in the CCW direction in quadrant III. When the machine works as a generator in the CW state, it is in quadrant IV (positive voltage and negative current). Quadrant II, negative voltage and positive current, is the CCW generator region.

When the H-bridge is operated in the three-level mode, four-quadrant operation is again available.

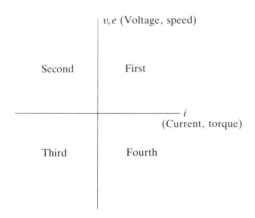

Fig. 9.19. Four-quadrant operation of an H-bridge.

10 Position control using a direct-current motor

This chapter is concerned with the fundamental theory, system hardware, and position control software of d.c. servomotors. First, a positioning system that uses both analogue and digital techniques is presented, and the roles of the microcomputer as the position-error counter as well as of the speed-command generator are discussed. A theory will be developed from both dynamical and electrical points of view. From this theory, a concept for determining speed commands as a function of the position error will be derived. The microprocessor used here is an Intel eight-bit 8085A, and an assembly-language program produced for executing control is explained in detail. This chapter is a revised translation of a part of author's Japanese book.[1]

10.1 System outline

Figure 10.1 shows the block diagram of a position-control system using a conventional d.c. motor. This is a simple but useful example for studying the

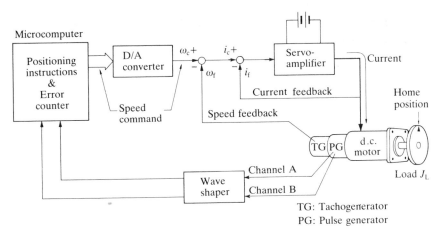

Fig. 10.1. Positioning servomechanism using a d.c. motor, a pulse generator (PG), a tachogenerator (TG), and a 8085 microprocessor as the position-error counter. The load of the motor is here an inertial load. In many applications, however, the load is driven via gears or pulleys.

Fig. 10.2. A laboratory experimental set-up.

practice of the theory of d.c. servomotors and application of microcomputers. The system can be realized using laboratory components as shown in Fig. 10.2, and consists of the following components:

(1) permanent-magnet d.c. servomotor;
(2) pulse generator (PG) coupled to the motor shaft;
(3) tachogenerator (TG) coupled to the motor shaft;
(4) servo-amplifier of the current-control type;
(5) error counter using a microprocessor (the one-board microprocessor in the MECHATRO LAB is used here);
(6) D/A converter;
(7) power supply.

A remarkable feature of this system is that the function of error counter is performed by a microcomputer. One of the major objectives of this chapter is to explain what the error counter is and how the microprocessor is utilized for this purpose.

As the servo-amplifier, either a linear or a PWM type can be used. Whichever type is chosen, however, the current-controlled scheme is preferred to the voltage-controlled one. The current feedback loop in the system block diagram in Fig. 10.1 corresponds to the current-detection signal loop discussed in Sections 5.3 and 5.4.2.

10.2 Theoretical basis

In a positioning system employing feedback control of a servomotor, the control of motor speed in acceleration/deceleration and slewing plays an important role. It is therefore helpful first to discuss the theoretical background between speed control and position control.

10.2.1 Rotational speed and position

The relation between the rotor position θ, which here is the positional angle, and the rotational speed ω can be expressed by

$$\theta = \int \omega \, dt. \tag{10.1}$$

This means that the integral of speed ω with respect to time is the rotational angle. Hence, for controlling θ we need to control ω.

10.2.2 Rules of dynamics

When torque T is applied to the rotor, the relationship between T and ω is given by the Newtonian dynamic equation:

$$(J_M + J_L) \frac{d\omega}{dt} + D\omega = T, \tag{10.2}$$

where J_M = moment of inertia of rotor
J_L = moment of inertia of load
D = viscous damping coefficient, and
T = torque generated by motor.

This means that for controlling speed in acceleration/deceleration, we need to control the torque. Let us examine this matter in detail.

(1) *Acceleration.* A motor is usually accelerated by a constant torque T generated by the maximum allowable current. The solution of eqn (10.2) is given by

$$\omega = \frac{T}{D} \left[1 - \exp\left(\frac{-Dt}{J_M + J_L}\right) \right]. \tag{10.3}$$

The graph of this function is shown in Fig. 10.3.

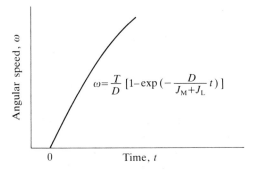

Fig. 10.3. Acceleration curve.

(2) *Deceleration.* By reversing the current to make the torque the negative maximum, the motor can effectively be decelerated. It is usual to ignore the term $D\omega$ and the dynamic equation is written as

$$(J_M + J_L)\frac{d\omega}{dt} = -T. \tag{10.4}$$

The solution of this equation is

$$\omega = \omega_{cs} - \frac{T}{J_M + J_L}\,t, \tag{10.5}$$

where ω_{cs} is the slewing speed.

(3) *Constant speed or slewing.* As the time differential term in eqn (10.2) is zero, the solution of this equation becomes as

$$\omega = \frac{T}{D}. \tag{10.6}$$

Since D is small in many applications, the torque T to maintain this speed can be low.

10.2.3 *Electric dynamics*

In a d.c. motor, the torque generated is proportional to the current i, and is expressible by the simple equation

$$T = K_T i, \tag{10.7}$$

where K_T is a parameter known as the torque constant and is different from motor to motor.

10.2.4 *Voltage equation*

To control current i, the voltage v applied to the motor must be controlled. The relationship between these two quantities is determined by

$$L\frac{di}{dt} + Ri + K_E\omega = v(t), \tag{10.8}$$

where R = armature resistance seen from the motor terminals
L = inductance seen from the motor terminals
K_E = back e.m.f. constant. (This is the same as the torque constant in physics. See Table 10.1. For details, see the Appendix of Reference [2].)

Table 10.1. Conversion table between torque constants and back e.m.f. constants.

Torque constant, K_T			Back e.m.f. constant, K_E	
N m/A	kg cm/A	oz in/A	V s/rad	V/kr.p.m.
1	10.2	141.6	1	104.7
0.09807	1	13.89	0.09807	10.27
0.007061	0.07200	1	0.007061	0.7394
0.009549	0.09738	1.352	0.009549	1

One problem among these relations is the last voltage equation. Since the equation is somewhat complex, it is seen that the current cannot be controlled simply by the voltage. For this reason, in many applications, the control system can be unstable. Hence, it is desirable that the current is directly controlled by the control signal. This is realized by a current feedback loop in the servomechanism. This type of servo-amplifier is known as the current-controlled servo-amplifier and was discussed in Section 5.3.

10.3 Speed profile for position control

In Fig. 10.4 a speed profile from start to stop at the target position is shown. This profile can be divided into five regions: acceleration, slewing, deceleration, settling, and stop. The functions in each region are as follows.

(1) *Acceleration.* The error counter instructs a maximum speed value towards which the rotor is accelerated according to eqn (10.3). The torque in this case is given by eqn (10.7) by using the maximum allowable current for i. Here, the maximum allowable current is the upper limit of the current, which is set to protect the solid-state power devices used in the servo-amplifier. The automatic setting mechanism is usually incorporated inside the amplifier.

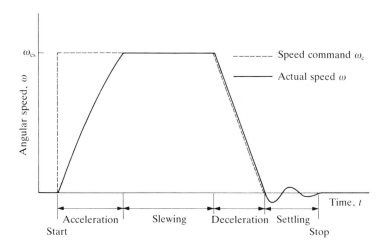

Fig. 10.4. Speed profile in position control.

During acceleration, the speed-feedback loop shown in Fig. 10.1 exerts no effect.

(2) *Slewing.* When the rotor speed reaches the maximum value, speed-control feedback is effective. As will be clarified later, the maximum speed is instructed by digital 127 or -128 and converted to analogue 5 or -5 V. In this experimental system, the motor is set so that it rotates at 1000 r.p.m. ($= 105$ radians/s) at this point.

(3) *Deceleration.* To decelerate the rotor according to eqn (10.7), the speed instruction must be given by

$$\omega_c = (2q\beta)^{\frac{1}{2}}, \tag{10.9}$$

where q = error (distance to the target value from the current position in terms of steps or number of pulses counted by the pulse generator mounted on the shaft)

β = rate of deceleration or negative acceleration.

Let us see what is implied in this important relation. First, from eqns (10.7) and (10.9), deceleration β is derived as

$$\beta = \frac{K_T i}{J_M + J_L}. \tag{10.10}$$

For the electric current i appearing in this equation, one can use a value that is appropriate for the safety of the drive of power devices used in the servo-

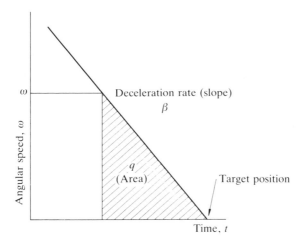

Fig. 10.5. Relationship between speed ω, deceleration rate β, and position error q.

amplifier and that will not demagnetize the permanent magnet used in the motor. In this section, however, another principle is employed. Before this is discussed, the derivation of eqn (10.9) should be explained.

Referring to Fig. 10.5, when the rotor is being decelerated, rotational speed ω at time t becomes

$$\omega = (t_0 - t)\beta, \tag{10.11}$$

where t_0 = time reaching the target position. The distance to the target position is given, using the formula for area of a triangle, by

$$q = \tfrac{1}{2}(t_0 - t)\beta(t_0 - t). \tag{10.12}$$

Eliminating $(t_0 - t)$ from both equations, we obtain the following equation.

$$\omega = (2q\beta)^{\frac{1}{2}}. \tag{10.13}$$

When ω is replaced with the command value ω_c for speed, this equation is identical to eqn (10.9). When deceleration is performed according to the eqn (10.13), the rotor stops rotation exactly at the target position.

The selection of β is again discussed here. As stated in the description for slewing (2), we set the speed at 1000 r.p.m. (or 105 radians/s) when digital 127 is instructed as the speed command. This means that digital 127 is sent out as the speed instruction when the position error is positive 127 steps or more. Likewise, when the error is -128 steps or larger, the output value is digital -128. With this arrangement, the distance travelled from initiating deceleration until the first arrival at the target is 127 (or 128) steps.

When a pulse generator with 200 steps per revolution is used, this distance is about 4.02 radians. Hence, by substituting $\omega = 105$ radians/s and $q = 4.02$ in eqn (10.13), the rate of deceleration is obtained as

$$\beta = 105^2/(2 \times 4.02) = 1.37 \times 10^3 \text{ radians/s}^2. \qquad (10.14)$$

If this value is smaller than the value of β given by eqn (10.10), acceleration is achievable. For example, if the maximum torque, which is $K_T i$ in eqn (10.10), is 2 N m and the total inertia is 1.3×10^3 kg m^2, the possible acceleration is 1.54×10^3 radians/s^2.

The time spent before stopping from the speed of 1000 r.p.m. is

$$105/(1.37 \times 10^3) = 0.0766 \text{ s} = 76.6 \text{ ms}. \qquad (10.15)$$

(4) *Settling.* Ideally, the motor stops at the target position and does not travel any farther when control is governed exactly by eqn (10.13). In practice, however, there is always a possibility that the rotor overruns beyond the target, because the speed command data are not given as an ideal linear function but discontinuously owing to digital quantization. Normally, the system is arranged so that the rotor eventually settles after several cycles of oscillation around the target. The time from the first passage of the target to the instant beyond which the rotor always stays within an allowable limit is referred to as settling time.

(5) *Stop.* When the rotor comes to rest at the target, the speed-feedback loop is usually cut off and the position-feedback control is initiated to yield a strong holding torque. If an error of ± 1 step is allowable at the target position, which is often the case, this change-over to position feedback is not always necessary.

10.4 Generation of speed commands and function of the error counter

In this instance, an 8085A microprocessor is used in the portion where the positional error is counted and the speed instruction is generated. The block diagram of Fig. 10.6 illustrates how the generation of the speed command is implemented.

10.4.1 *Method of providing positional instruction data*

As shown in the block diagram of Fig. 10.1, a disc with a reference mark is mounted on the motor shaft. When a certain reference position is initially defined, the distance the motor has travelled from it can be expressed in terms of the number of steps or pulses generated from the pulse generator. The positional instruction data can also be expressed by the number of steps to travel and the direction code.

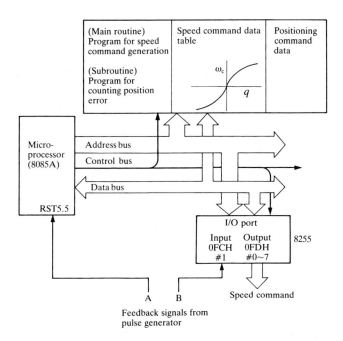

Fig. 10.6. Block diagram around the microprocessor.

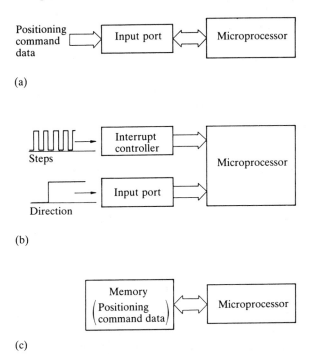

Fig. 10.7. Methods of sending positioning data: (a) parallel data; (b) serial data; (c) storage in memory area.

Three methods of providing instruction data are illustrated in Fig. 10.7:

(1) Instruction is made by parallel data to be given from an input port.
(2) Instruction is made with pulse train by interrupt signals to the microprocessor.
(3) The data is accommodated in a certain memory area and read out in sequence.

Here we take method (3), because this is the simplest and is suited for beginners. As shown in Fig. 10.8, the instruction data, which indicate number of steps and direction (CW/CCW) to be travelled from the current position, is accommodated as signed 16-bit binary codes in the memory area from address 8210H. End of data is signified by zero.

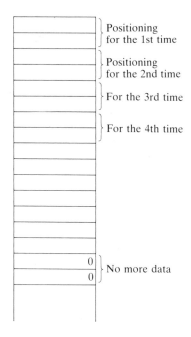

Fig. 10.8. Memory area storing the position-command data.

10.4.2 *Position feedback*

The feedback information on position should be given to the #1 bit of input port FC and RST5.5 terminal as shown in Fig. 10.9. The channel A signal is sent to the RST5.5 terminal of the microprocessor as an interrupt signal, and the channel B signal enters the input port. In this scheme, the rotor movement is detected as follows:

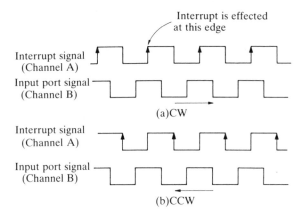

Fig. 10.9. Principle of discriminating rotational direction and counting position error using a two-channel signal from the pulse generator. The channel A signal is supplied to the interrupt terminal (RST5.5 here) and the channel B signal to the #1 bit of input port of address FCH. Here the interrupt request is received at a rising edge of the signal. Soon after the interrupt is received, the state of the #1 bit of port FCH is inspected by the interrupt subroutine. When this is H, it is taken that the motor has travelled one step in a CW direction, and the error data is incremented. When the level at the #1 bit is L, it is taken that the motor has travelled one step in a CCW direction, and the error data is decremented.

(1) When an interrupt signal arrives, it means that a motor has travelled one step.
(2) At this time, if the #1 bit of input FC is at the H level, it is taken that the rotation has been clockwise (CW); if it is at the L level, it is taken as CCW.

10.4.3 *Main routine (generation of speed instruction)*

The main routine is a program to generate the speed instruction as a function of the position error, which is the distance to the target from the current position. The output is produced according to the following rules.

(1) For errors (q) from zero to 127 steps, the speed command should be a digital value approximation computed using

$$\omega_c' = (127q)^{\frac{1}{2}}. \qquad (10.16)$$

(2) For errors from 0 to -128 steps,

$$\omega_c' = -(-128q)^{\frac{1}{2}}. \qquad (10.17)$$

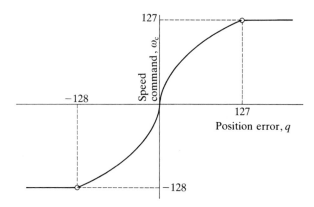

Fig. 10.10. Relationship between the digital values of speed commands and position errors.

(3) For errors from 128 to 32767 steps,

$$\omega_c' = 127. \tag{10.18}$$

(4) For errors from -129 to -32768,

$$\omega_c' = -128. \tag{10.19}$$

Figure 10.10 shows the graph of this relation. The positive maximum instruction for the speed is 127, while it is -128 in the opposite speed. These values are only selected for ease of treatment of signed eight-bit numbers. Data 127 and 128 are considered to be almost the same as the maximum speed command data.

Caution is necessary regarding the unit of the speed command. In the theory in Section 10.3, the unit of the speed and its command is radians per second. However, in eqns (10.16) to (10.19), the ω_c' are non-dimensional numbers. The relations between these are determined by characteristics of the D/A converter and the tachogenerator coupled to the motor. In this example, the D/A converter and the tachogenerator are adjusted so that the digital 127 corresponds to 1000 r.p.m. or 105 radians/s. As stated in Section 10.3 under (2), when 127 is instructed the output voltage from the D/A converter is 5 V and when the speed is 1000 r.p.m. the tacho output is also 5 V in this example. For more details, refer to Sections 10.5.2 and 10.5.4.

The flowchart of the program for generating speed instructions according with this rule is shown in Figs. 10.11 and 10.12. A detailed explanation will be given later. However, it is desirable to clarify here how to deal with the computation of speed commands as a function of the error between -128 and 127. The square roots are not calculated in the program: they should be

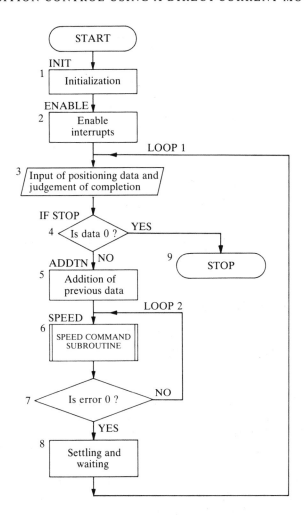

Fig. 10.11. Flowchart of the main routine.

calculated beforehand, and the results stored in the memory area 8100H to 81FFH as shown in Table 10.2, so that the required value is always available to be read out.

It is important that negative numbers appear in the speed and position instruction data (or positional errors). In Fig. 10.13 two different formats for 8-bit signed data are compared. Figure 10.13(a) shows the two's complement expression, and (b) shows the offset binary expression. The difference of these two is seen in the interpretation of the most significant bit or MSB. The former is convenient for arithmetic operation in a microprocessor, and

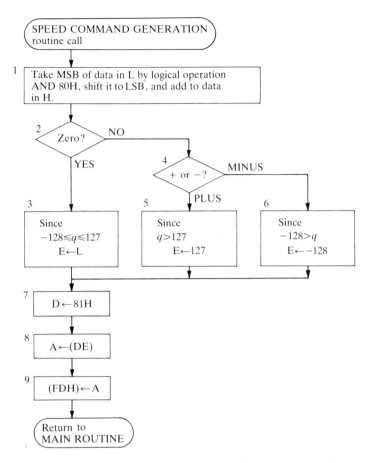

Fig. 10.12. Flowchart of the speed-command-generation routine.

the latter is suitable as the digital data to be supplied to a D/A converter. The speed instruction data is, hence, given in the offset binary format. Sixteen-bit offset binary codes are expressed according to the rule explained in Fig. 10.14.

The speed command data table is shown in Table 10.2, which uses a memory area from 8100H to 81FFH. This is found in the program list of Table 10.3 under ACCEL DECEL DATA TABLE.

10.4.4 *Interrupt subroutine (error counter)*

The interrupt subroutine is a program to receive a signal from the pulse generator and compute the error. The flowchart of this part is simple, as shown in

Table 10.2. Look-up table giving the relation between the position errors and speed instruction data.

Position error (decimal)	Address (hexadecimal)	Speed command data (decimal)
0	8100	0
1	8101	11
2	8102	16
3	8103	20
4	8104	23
5	8105	25
6	8106	28
7	8107	30
8	8108	32
9	8109	34
10	810A	36
124	817C	125
125	817D	126
126	817E	126
127	817F	127
− 128	8180	− 128
− 127	8181	− 127
− 126	8182	− 127
− 125	8183	− 126
− 124	8184	− 126
− 123	8185	− 125
− 122	8186	− 125
− 5	81FB	− 25
− 4	81FC	− 23
− 3	81FD	− 20
− 2	81FE	− 16
− 1	81FF	− 11

Note. Lower bytes of the addresses designate the position errors in two's complement. In Table 10.3, the speed commands are given in the offset binary expression.

Fig. 10.15. What should be noted here is that pair register HL is used for computing the error; the data in pair register HL always indicate the distance to the target in terms of number of steps. The details of algorithm are explained in Section 10.6.

10.5 Hardware supplement

Before a detailed explanation of software, we give a simple supplementary explanation of the elements involved in the systems.

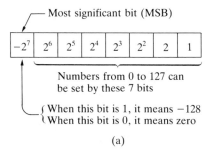

(a)

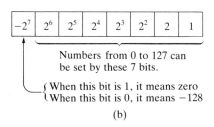

(b)

Fig. 10.13. Two formats for signed eight-bit data: (a) two's complement format; (b) offset binary format.

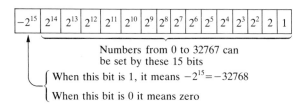

Fig. 10.14. Two's complement format for signed 16-bit data.

Table 10.3. Program listing for position control using a d.c. servomotor.

```
;**** POSITION CONTROL USING A DC MOTOR ****

        ORG     8000H

DRIVE   EQU     0FDH      ;Output port
ENC     EQU     0FCH      ;Input port
CMD     EQU     0FFH      ;Command port
IRADD   EQU     83C0H     ;RST5.5 Address
STACK   EQU     83FFH     ;Initial stack pointer
FACTOR  EQU     200       ;Waiting time factor

INIT:   DI                ;Disenable interrupts
        LXI     SP,STACK  ;Set stack pointer
        MVI     A,90H     ;I/O port command data
        OUT     CMD       ;Set I/O ports
        MVI     A,80H     ;Load A with 80H(=offset binary zero)
        OUT     DRIVE     ;Output to DRIVE
        MVI     A,0C3H    ;Set starting address for
        LXI     H,INT5R5  ;RST5.5 POSITION ERROR COUNTER ROUTINE
        STA     IRADD     ;Store RST 5.5 address in A
        SHLD    IRADD+1   ;Store next address in HL
        LXI     H,0       ;Load H with initial error zero
        LXI     D,DATA    ;Load DE with starting address for
                          ;position command data
        PUSH    D         ;Push DE down to stack

ENABLE: EI                ;Enable interrupts

LOOP1:  POP     D         ;Retrieve position command address
        LDAX    D         ;Fetch lower byte data
        MOV     C,A       ;and store it in C
        INX     D         ;Increment command address
        LDAX    D         ;Fetch higher byte data
        MOV     B,A       ;and store it in B
        INX     D         ;Increment command address
        PUSH    D         ;Push DE down to stack

IFSTOP: MOV     A,B       ;
        ORA     C         ;Logical OR of A(=B) and C
        JZ      STOP      ;If BC=0 go to STOP

ADDTN:  DAD     B         ;Add new command data to previous
                          ;data remaining in HL

LOOP2:  CALL    SPEED     ;Go to SPEED COMMAND SUBROUTINE
        MOV     A,H       ;
        ORA     L         ;Logical OR of A(=H) and L
        JNZ     LOOP2     ;If position error is not zero
                          ;then go to LOOP2
```

Table 10.3. *continued*

```
;***** SETTLING/WAITING ROUTINE *****

TIME:    MVI    B,FACTOR  ;(Wating time) Load B with FACTOR
LOOP3:   MVI    C,0       ;Load C with zero
SETTLE:  CALL   SPEED     ;Settling
         DCR    C         ;Decrement C
         JNZ    SETTLE    ;If not zero go to SETTLE
         DCR    B         ;Decrement B
         JNZ    LOOP3     ;If not zero go to LOOP3
         JMP    LOOP1     ;Go to LOOP1

;***** STOP COMMAND EXECUTION *****

STOP:    MVI    A,80H     ;Load A with 80H(=offset binary zero)
         OUT    DRIVE     ;Output to DRIVE
         DI               ;Disenable interrupts
         HLT              ;Halt CPU

;***** SPEED COMMAND SUBROUTINE *****

SPEED:   MOV    A,L       ;Check whether postion error is
         ANI    80H       ;between -128 and 127, or over
         RLC              ;this range
         ADD    H         ;
         JNZ    OVER      ;If not zero, then go to OVER
         MOV    E,L       ;Since -128 < position error < 127
                          ;move data from L to E
         JMP    OUTPUT    ;Go to OUTPUT
OVER:    MOV    A,H       ;Check whether position error is
         ANA    A         ;negative or positive
         JM     MINUS     ;If negative go to MINUS
PLUS:    MVI    E,127     ;As positive, load E with 127
         JMP    OUTPUT    ;Go to OUTPUT
MINUS:   MVI    E,-128    ;As negative, load E with -128
OUTPUT:  MVI    D,81H     ;Load D with 81H
         LDAX   D         ;Load A with data stored in (DE)
         OUT    DRIVE     ;Output to D
         RET              ;Return to main routine

;***** RST5.5 POSITION ERROR COUNTER ROUTINE *****

INT5R5:  PUSH   PSW       ;Push flag state down to stack
         IN     ENC       ;Fetch encoder signal
         JZ     CCW       ;If zero, go to CCW

CW:      INX    H         ;Increment HL
         POP    PSW       ;Retrieve flag state
         EI               ;Enable interrupts
         RET              ;Return to main routine

CCW:     DCX    H         ;Decrement HL
         POP    PSW       ;
         EI               ;
         RET              ;
```

Table 10.3. *continued*

```
;***** ACCEL/DECEL DATA TABLE *****

        ORG     8100H

TABLE:  DB      080H,08BH,08FH,093H,096H,099H,09BH,09DH
        DB      09FH,0A1H,0A3H,0A5H,0A7H,0A8H,0AAH,0ABH
        DB      0ADH,0AEH,0AFH,0B1H,0B2H,0B3H,0B4H,0B6H
        DB      0B7H,0B8H,0B9H,0BAH,0BBH,0BCH,0BDH,0BEH
        DB      0BFH,0C0H,0C1H,0C2H,0C3H,0C4H,0C5H,0C6H
        DB      0C7H,0C8H,0C9H,0C9H,0CAH,0CBH,0CCH,0CDH
        DB      0CEH,0CEH,0CFH,0D0H,0D1H,0D2H,0D2H,0D3H
        DB      0D4H,0D5H,0D5H,0D6H,0D7H,0D8H,0D8H,0D9H
        DB      0DAH,0DAH,0DBH,0DCH,0DCH,0DDH,0DEH,0DEH
        DB      0DFH,0E0H,0E0H,0E1H,0E2H,0E2H,0E3H,0E4H
        DB      0E4H,0E5H,0E6H,0E6H,0E7H,0E7H,0E8H,0E9H
        DB      0E9H,0EAH,0EAH,0EBH,0ECH,0ECH,0EDH,0EDH
        DB      0EEH,0EEH,0EFH,0F0H,0F0H,0F1H,0F1H,0F2H
        DB      0F2H,0F3H,0F4H,0F4H,0F5H,0F5H,0F6H,0F6H
        DB      0F7H,0F7H,0F8H,0F8H,0F9H,0F9H,0FAH,0FAH
        DB      0FBH,0FBH,0FCH,0FCH,0FDH,0FDH,0FEH,0FFH
        DB      000H,001H,001H,002H,002H,003H,003H,004H
        DB      004H,005H,005H,006H,006H,007H,007H,008H
        DB      008H,009H,009H,00AH,00AH,00BH,00BH,00CH
        DB      00DH,00DH,00EH,00EH,00FH,00FH,010H,011H
        DB      011H,012H,012H,013H,013H,014H,015H,015H
        DB      016H,016H,017H,018H,018H,019H,019H,01AH
        DB      01BH,01BH,01CH,01DH,01DH,01EH,01FH,01FH
        DB      020H,021H,021H,022H,023H,023H,024H,025H
        DB      025H,026H,027H,027H,028H,029H,02AH,02AH
        DB      02BH,02CH,02DH,02DH,02EH,02FH,030H,031H
        DB      031H,032H,033H,034H,035H,036H,036H,037H
        DB      038H,039H,03AH,03BH,03CH,03DH,03EH,03FH
        DB      040H,041H,042H,043H,044H,045H,046H,047H
        DB      048H,049H,04BH,04CH,04DH,04EH,050H,051H
        DB      052H,054H,055H,057H,058H,05AH,05CH,05EH
        DB      060H,062H,064H,066H,069H,06CH,070H,074H

;***** POSITION COMMNADS DATA *****

        ORG     8200H

DATA:   DW      100
        DW      -200
        DW      300
        DW      -400
        DW      500
        DW      -600
        DW      700
        DW      -800
        DW      900
        DW      -1000
        DW      0

        END
```

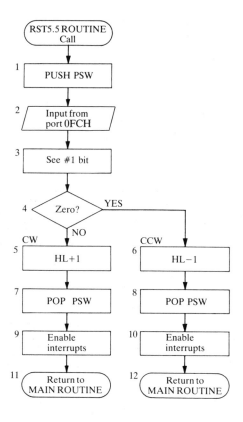

Fig. 10.15. Flowchart of the position-error counter routine.

10.5.1 *I/O port*

This system uses an Intel 8255A-5 chip under mode zero as the parallel I/O device and the signal for each port is as follows:

Input port address: 0FCH
#1bit: Channel B signal of the pulse generator
Other bits: Not used
Output port address: 0FDH
#0 to 7 bits: speed instruction (-128 to 127).

10.5.2 *D/A converter*

The outline of how to use a D/A converter is shown in Fig. 10.16. Figure 10.16(a) shows the circuit employed in this system using a popular D/A con-

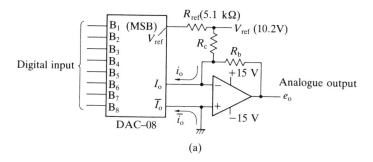

(a)

Decimal	Digital input values								Output current	
	B_1	B_2	B_3	B_4	B_5	B_6	B_7	B_8	$i_o(\mu A)$	$\overline{i}_o(\mu A)$
127	1	1	1	1	1	1	1	1	1992	0
126	1	1	1	1	1	1	1	0	1984	8
125	1	1	1	1	1	1	0	1	1976	16
3	1	0	0	0	0	0	1	1	1023	968
2	1	0	0	0	0	0	1	0	1016	976
1	1	0	0	0	0	0	0	1	1008	984
0	1	0	0	0	0	0	0	0	1000	992
−1	0	1	1	1	1	1	1	1	992	1000
−2	0	1	1	1	1	1	1	0	984	1008
−3	0	1	1	1	1	1	0	1	976	1016
−126	0	0	0	0	0	0	1	0	16	1976
−127	0	0	0	0	0	0	0	1	8	1984
−128	0	0	0	0	0	0	0	0	0	1992

(b)

Fig. 10.16. D/A converter: (a) using an operational amplifier and arrangements of resistors; (b) relation between the input digital values and the output currents.

verter IC, such as DAC–108. In using such a device the potential applied to the V_{ref} terminal and the resistance R_{ref} are very important. In this example the reference potential is 10.2 V and R_{ref} is 5.1 kΩ. The so-called I_{ref} value in this case is 10.2/5.1 k = 2 mA.

In this device, the maximum sink current at terminal I_o is determined by

$$I_o = (255/256) \times 2000 \ \mu A = 1992 \ \mu A, \qquad (10.20)$$

and the overall relationships between the input digital values and the output currents are as shown in Fig. 10.16(b). In the circuit configuration of (a), however, the current flowing to the $\bar{I}_o$ terminal has no influence to the output voltage e_o, because the $(+)$ terminal of the operational amplifier is grounded.

Now let us derive the relation between the digital input quantities and the output analogue voltages e_o. In deriving this relation we should pay attention to the following two characteristics of the operational amplifier.

(1) The potential difference between the $(+)$ and $(-)$ terminals are negligibly small. Since in this application the $(-)$ terminal is grounded, the voltage at the $(+)$ terminals must always be zero.
(2) Since the input impedance is infinite, no current can enter either the $(+)$ or $(-)$ terminals.

The current flowing towards the $(-)$ terminal of the operational amplifier consists of three components. One is $-i_o$ coming from the D–A converter, the second is V_{ref}/R_c which is supplied from the reference voltage supply via R_c, and the last is e_o/R_b or the fed-back current from the output terminal. Owing to item (2), the summation of these three must be zero:

$$-i_0 + \frac{V_{ref}}{R_c} + \frac{e_0}{R_b} = 0, \tag{10.21}$$

from which we get

$$e_0 = i_0 R_b - \frac{R_b}{R_c} V_{ref}. \tag{10.22}$$

According to the table in Fig. 10.16(b), when the digital input is 0, $i_0 = 1000\ \mu A$, and hence e_0 becomes

$$e_0 = R_b \left(10^{-3} - \frac{V_{ref}}{R_c} \right). \tag{10.23}$$

To make this analogue value zero, V_{ref}/R_c must be chosen to be 10^{-3} A. Since in this example V_{ref} is 10.2 V, R_c must be 10.2 kΩ.

The minimum output, which occurs when digital -128 is applied to the input terminal, is

$$(e_0)_{min} = -\frac{R_b}{R_c} V_{ref}. \tag{10.24}$$

If R_b is 5 kΩ, the value of $(e_o)_{min}$ is -5 V. On the other hand, when the maximum 127 is applied to the input terminal, the output potential is

$$(e_0)_{max} = 5 \times 10^3 (1992 \times 10^{-6} - 10^{-3}) = 4.96\ \text{V}. \tag{10.25}$$

10.5.3 Direct-current servomotor

The motor used here is the type having a normal rotor. In this system, a tachogenerator (TG) and pulse generator (PG) are coupled directly to the shaft.

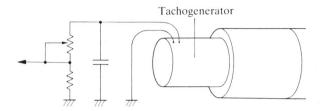

Tachogenerator

Fig. 10.17. Adjusting the feedback voltage from the tachogenerator using a rheostat.

10.5.4 *Direct-current tachogenerator*

The tachogenerator is a kind of d.c. generator that converts the rotational speed value into a d.c. voltage. Let us suppose that the rated output voltage is 7 V when the tachogenerator rotates at 1000 r.p.m. By shunting the output voltages using a rheostat as shown in Fig. 10.17, the voltage fed back to be compared to the D–A converter output is adjusted to 5 V for the given speed of 1000 r.p.m.

10.5.5 *Pulse generator*

A rotary encoder or pulse generator is used to detect the amount of rotor movement and its direction. There must be a phase difference of about 90° between the signals from channels A and B so that the direction of motion can be detected. The larger the split number of the slit disc, the higher the resolution in positioning. However, in the position control method employed here, the encoder signals are directly processed by the microprocessor, and hence the split number N must be selected to satisfy the following relationship:

$$N \leqslant \frac{600\,000}{\text{Maximum rotational speed in r.p.m.}}. \tag{10.26}$$

The upper limit of the interrupting frequency is here assumed to be 10 kHz. The calculation of this datum is a little complex. First, it is related to the time from the start of the error counter subroutine, which is initiated by the application of an interrupt, to the return to the main routine after completion of its processing. It is desirable to make the time spent in this routine as short as possible. In the program to be explained later, it becomes about 40 μs for a clock frequency of 2.5 MHz. Hence, the maximum interrupt frequency can be as high as 25 kHz when this is the only consideration. However, if so, since the time to process the main routine is not available, we desire the maximum interrupting frequency to be as low as 10 kHz. Since the maximum speed of this system is 1000 r.p.m. or 16.7 r.p.s., the maximum number of pulses per rotation is 600. Here, a pulse generator having a resolution of 200 is selected.

10.5.6 *Wave shaper of the pulse generator signals*

Signals from the pulse generator mounted on the motor are fed to the circuit shown in Fig. 10.18 before being supplied to the input port and the RST5.5 terminal of the microprocessor. Via two logic inverters, the signal from channel B goes to the #1 bit of the input port whose address is, for example, 0FCH. Instead of these inverters, one can use a single buffer. Interruption should be effected at the build-up edges of channel A signal. Using an RC delay circuit and an AND gate, a pulse train is generated from the signals of channel A and fed to the RST5.5 terminal.

10.6 Explanation of the program

We have reached the stage of explanation of the program based on the fore-going considerations. The program listing in Table 10.3 and the flowcharts of Figs. 10.11, 10.12, 10.15 and 10.19 will be referred to.

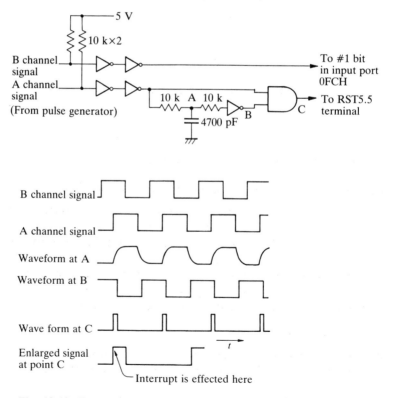

Fig. 10.18. Example of circuit for processing pulse-generator signals.

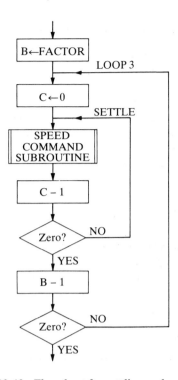

Fig. 10.19. Flowchart for settling and waiting.

10.6.1 *Roles of registers and use of I/O ports*

The registers of the 8085A microprocessor are used as follows.

Pair register BC: (1) temporary memory used when taking positional instruction data, and

                    (2) counters in the time routine as separate two registers

Pair register DE: (1) addressing positioning command data, and

                    (2) addressing speed command data

Pair resister HL: counting and memory of position error.

The roles of I/O ports are as follows.

Address 0FDH (Label Drive):   Output port for speed instruction

Address 0FCH (Label ENC):   Input port for the B channel signal (#1 bit)

Address 0FFH (Label CMD):   Control port of the I/O chip.

The memory allocation is as follows.

(1) From address 8000H the main routine, which is the generation of speed instruction, is written.

(2) From address 80B0H the interrupting subroutine or the error counter routine is written.

(3) From 8100H to 81FFH—speed instruction table.

(4) From 8210H—positioning instruction data.

10.6.2 *Error counter*

Before discussing the main-routine program, the interrupt-subroutine program should be explained. This is a relatively simple routine for counting the position error. The flowchart is shown in Fig. 10.15, and the program listing of 15 lines is seen under RST5.5 POSITION ERROR COUNTER ROUTINE in Table 10.3. Here, the explanation is given in the order of block numbers of the flowchart.

(1) *PUSH PSW.* The content of registers A and PSW (processor status word or flag register) are pushed down to the stack area.

(2) *Input from port 0FCH (Label ENC).* The data is taken from port 0FCH. Channel B signal is supplied to the #1 bit of this port.

(3) *See #1 bit.* By taking the logical product (AND) with 02H, the data of the #1 bit is picked up.

(4) *Zero?.* Here, the rotational direction is checked by the zero flag. As explained in Fig. 10.14, if this is zero it is judged to be positive and to be for a CCW movement and if 1 the datum is negative and is for a CW movement.

(5) *and* (6) *HL + 1 and HL − 1, respectively.* This is the position-error-counting routine. Before this routine is executed, the initial positioning instruction data should be loaded in pair register HL. As this routine is executed, the absolute value of this datum will be reduced rapidly. That is, when the movement is CCW, the position error must be positive and decremented as one step travels approaching the target. When movement is CW, the error is negative the data is incremented as the motor travels one step angle. This is the meaning of counting the position error.

(7) *and* (8) *POP PSW.* Since computation of the position error has been completed, the contents of register A and the flag (or processor status word) register are pushed down in the stack area.

(9) *and* (10) *Enable interrupts.*   Before the program returns to the main routine, an 'Enable Interrupts' instruction is given, because a 'Disenable Interrupts' instruction is automatically set when an RST5.5 interrupt signal is applied.

(11) *and* (12) *Return to MAIN ROUTINE.*   The program returns to the main routine.

10.6.3 *Main-routine program*

The intial part of the program listing is the main routine. 'ORG 8000H' of the first line is the declaration of the starting address and the next three lines are the declaration of the labels of the I/O ports.

The output port used to output the speed command data to the D–A converter is labelled DRIVE. The label for the input port receiving channel B signal from the pulse generator mounted on the motor is ENC. The command address to the I/O port is referred to by the label CMD. The correspondence between the labels and the real addresses is

DRIVE	EQU	0FDH
ENC	EQU	0FCH
CMD	EQU	0FFH

The labels for interrupt address and initial stack pointer are

IRADRS	EQU	83C0H
STK	EQU	8400H

The actual program starts thereafter, and its flowchart is shown in Fig. 10.11. Explanation is given in the order of the blocks.

(1) *Initialization (INIT).*   Before running the program as scheduled, the initial data of the registers should be set and use of the I/O ports and purpose of the memory area must be instructed. This process is called initialization and is explained in details as follows.

(i) *Disenabling interrupts.*   It is required to disenable interrupts until the initialization is completed.

(ii) *Specifying stack pointer.*   The stack area is the memory area where the data of registers are pushed down when the PUSH command is executed. This address is to be declared at the start. The start address of the stack area, which is labelled STK, is set at 8400H here, and this address should be written (or declared) in the stack pointer. The declaration of the stack pointer is made by LXI SP, STK. The data in the stack can be reloaded in registers by a POP command.

(iii) *Setting mode of I/O port.* An 8255 chip has three eight-bit ports called A, B, and C. We shall use ports in mode zero, which is the simplest way. Here port A, whose address is 0FCH, is used as an input port, and ports B and C, whose addresses are 0FDH and 0FEH, respectively, are set as output ports. Port C is not used in this example. The control port labelled CMD is used to take instructions on how to use ports A, B, and C. The real address depends on the hardware, but in this example it is 0FFH.

Initialization of the 8255 chip is implemented by

$$\begin{array}{ll} \text{MVI} & \text{A,90H} \\ \text{OUT} & \text{CMD} \end{array}$$

(iv) *Instruction of zero speed.* Offset binary zero (80H in hexadecimal notation) is commanded through the port DRIVE, whose address is FDH, to set the motor speed to zero initially. The program is executed by

$$\begin{array}{ll} \text{MVI} & \text{A,80H} \\ \text{OUT} & \text{DRIVE} \end{array}$$

(v) *Specification of jump address at interrupting.* In the explanation given for the interrupt subroutine in Section 10.6.2, it was stated that the start address of this routine is 80B0H. It is necessary that the start address be set to this before the first interrupt signal comes to the RST5.5 terminal. In the 8085 microprocessor, when an interrupt of RST5.5 level is applied, in other words when the H level signal enters in the RST5.5 terminal, the program so far executed is interrupted and the program counter (PC) register is loaded with 002CH. This means that the program is compelled to jump to address 002CH. Hence, if an instruction 'to re-jump to 80B0H' has been written in the three bytes from 002CH, the program will jump again to 80B0H. This is instructed by the following four lines:

$$\begin{array}{ll} \text{MVI} & \text{A,0C3H} \\ \text{LXI} & \text{H,INT5R5} \\ \text{STA} & \text{IRADRS} \\ \text{SHLD} & \text{IRADRS}+1 \end{array}$$

(vi) *Clearing pair register HL.* In order that pair register HL is used as the error counter, the data must be cleared by the command

$$\begin{array}{ll} \text{LXI} & \text{H,0} \end{array}$$

(vii) *Specifying position-command data address.* The first address of the memory area storing the position-command data is labelled DATA. Pair register DE is loaded with this data and pushed down to the stack area, because this pair register will be used for another purpose, to specify the address of the speed command data later. These procedures are instructed by

LXI   D,DATA
PUSH   D

(2) *Enable interrupts.*   When the initialization is completed as above and the system is ready to begin executing positioning control, interrupt enable (EI) is instructed.

(3) *Input of positioning data and judgement of completion (LOOP1).*   The address of the position instruction data is loaded in pair register DE by POP D. The content of this address is transferred to register A by LDAX D and further transferred to register C by MOV C,A. Then the statements from INX D to MOV B,A transfer the data of next address to register B. In this manner, the initial position instruction is loaded into register BC. Then, the next address is accommodated in the stack area by INX D and PUSH D. LXI D, TABLE is the command to load the initial address (8100H) of the speed-instruction table into pair register DE.

(4) *Is data 0 ?.*   When the data loaded in pair register BC is zero, there is no longer any position data, and the program jumps to address STOP. Whether the content of resister BC is zero or not can be found by taking the logical addition (OR) of the data in registers B and C. When this is zero, it means that the contents of B and C are zero at the same time and hence the 16-bit number in BC is zero.

(5) *Addition of previous data.*   Addition of the data in pair register BC and the residual error or the data in pair register HL, which is initially zero, does not seem to have any meaning. The addition process after the first cycle is absolutely necessary, because when the motor has been moved by an external force during the SETTLING/WAITING routine, which is explained shortly, the position error can be cumulative unless the remaining error is compensated in the new cycle. Adding the remaining error to the positioning data in the next cycle is a simple means of error compensation. This is executed by DAD B.

(6) *SPEED COMMAND subroutine.*   After the motor has started, the position error will vary with time. Speed-command data are generated in this subroutine as a function of position error. The details will be explained in Section 10.6.4.

(7) *Is error 0?*   Whether the position error is zero or not is tested. If the error is not zero, the program branches to LOOP2 again to execute the SPEED COMMAND subroutine. When the initial error is large, the SPEED

COMMAND subroutine must be repeated many times before the error becomes zero.

(8) *Settling and waiting.*   Once the position target has been reached, the routine for SETTLING/WAITING is to be executed. The details of this routine are presented in the flowchart of Fig. 10.19. To produce a waiting time, registers B and C are used. Register B is loaded with a datum designated by FACTOR, which is 200 in the present example. Register C is loaded with an appropriate number, which is zero in this example.

As shown in Fig. 10.19, the SPEED COMMAND subroutine is used also in settling and waiting. This means that the position error is always checked and feedback is effected during the waiting-time period. Theoretically, if the speed command is given according to eqn (10.13), the motor stops at the target position. However, owing to digitization approximations, there is a possibility that the target is overshot. Even if the motor has stopped successfully at the target, it can be caused to deviate from the target by an external force before the next command is given. In order that the position deviation from the target can always be eliminated, the SPEED COMMAND subroutine is used repeatedly in the manner already described.

When registers B and C are both decounted to zero, the first cycle or the positioning instructed by the first datum is complete and the execution jumps to LOOP1 to fetch the next position-command datum. Subsequently, the second, third, etc., positioning will be continued until positioning data of zero is arrived at.

(9) *STOP.*   When execution arrives here, the offset binary zero or 80H is first commanded as the speed instruction, and next HLT is instructed to halt the CPU.

10.6.4 *Details of speed-command generation*

The detailed flowchart of this subroutine is shown in Fig. 10.12. In this subroutine, the range of magnitude of the error is first examined to see whether it falls in the range − 128 to 127. The numbers in the range from 0 to 127 and that from − 128 to 0 have signed 16-bit codes, as follows:

0 to 127:	0000 0000 0000 0000	to	0000 0000 0111 1111
− 128 to − 1:	1111 1111 1000 0000	to	1111 1111 1111 1111
	H L		H L

When the datum in pair register HL falls in this range, the apparent result of addition of the MSB value in register L, which is 1 or 0, and the total datum in register H becomes zero. Otherwise the result is not zero. The logical and arithmetic operations are executed in block (1) in the flowchart and the judgement is in part (2). The corresponding part in the program list is

```
MOV   A, L
ANI   80H
RLC
ADD   H
JNZ   OVER;   If this is not zero, jump to address OVER
              since this range is exceeded. If zero, execute
              the next command.
```

Subsequent explanation will be given in the order of the blocks in the flowchart.

(3) *− 128 to 127*.   Since the content of pair register H L falls in the range − 128 to 127 if the result is zero, the content of register L is transferred to register E. Then preparation is provided for output of $(127q)^{1/2}$ or $-(128(-q))^{1/2}$, where q is the position error.

(4) *Sign of H (OVER)*.   When the content of pair register H L falls outside − 128 to 127, a test whether it is positive or negative is next performed. This can be tested by examining the MSB of register H with the following instructions:

```
MOV   A,H    ;   The content of H is transferred to register A
ANA   A      ;   Take logic AND with itself
JMP   MINUS; If the result is minus, jump to address
                 MINUS.
```

(5) *Larger than 127 (PLUS)*.   The error is judged to be positive, which means that it is larger than 127. Hence, according to the rule explained in Fig. 10.10, 127 or 7FH is loaded into register E as the upper limit of the speed-instruction datum.

(6) *Larger than − 128 and negative (MINUS)*.   The error is judged to be larger than − 128 and negative, and hence − 128 or 80H is loaded into register E to specify the lower byte of the address containing the speed-command datum.

(7) *D ← 81H*.   Register D is loaded with 81H to specify the higher byte of the address storing the speed commands. Register E has already been loaded with an appropriate datum to specify the lower byte of the memory address storing the speed-command datum.

(8) $A \leftarrow (DE)$.   The datum in the memory specified by pair register DE is transferred to register A to be sent out to the output port in the next block.

(9) *Output of speed instruction data (OUTPUT)*.   The speed-command datum now stored in register A is sent out to port DRIVE, whose address is 0FDH.

(10) *RETURN*.   Program execution returns to the main routine.

References

1. Kenjo, T. (1986). *Mechatronics controls using Z80/8085*. (In Japanese). Chapter 10. Sogo Electronics Publishing Company, Tokyo.
2. Kenjo, T. and Nagamori, S. (1985). *Permanent-magnet and brushless DC motors*. Oxford University Press, Oxford.

Index